Nitro Compounds and Their Derivatives in Organic Synthesis

Nitro Compounds and Their Derivatives in Organic Synthesis

Editor

Nagatoshi Nishiwaki

MDPI • Basel • Beijing • Wuhan • Barcelona • Belgrade • Manchester • Tokyo • Cluj • Tianjin

Editor
Nagatoshi Nishiwaki
Kochi University of Technology
Japan

Editorial Office
MDPI
St. Alban-Anlage 66
4052 Basel, Switzerland

This is a reprint of articles from the Special Issue published online in the open access journal *Molecules* (ISSN 1420-3049) (available at: https://www.mdpi.com/journal/molecules/special_issues/nitro_organic_synthesis).

For citation purposes, cite each article independently as indicated on the article page online and as indicated below:

LastName, A.A.; LastName, B.B.; LastName, C.C. Article Title. *Journal Name* **Year**, *Article Number*, Page Range.

ISBN 978-3-03943-148-9 (Hbk)
ISBN 978-3-03943-149-6 (PDF)

© 2020 by the authors. Articles in this book are Open Access and distributed under the Creative Commons Attribution (CC BY) license, which allows users to download, copy and build upon published articles, as long as the author and publisher are properly credited, which ensures maximum dissemination and a wider impact of our publications.

The book as a whole is distributed by MDPI under the terms and conditions of the Creative Commons license CC BY-NC-ND.

Contents

About the Editor vii

Nagatoshi Nishiwaki
A Walk through Recent Nitro Chemistry Advances
Reprinted from: *Molecules* **2020**, *25*, 3680, doi:10.3390/molecules25163680 1

Maxim A. Bastrakov, Alexey K. Fedorenko, Alexey M. Starosotnikov, Ivan V. Fedyanin and Vladimir A. Kokorekin
Synthesis and Facile Dearomatization of Highly Electrophilic Nitroisoxazolo[4,3-*b*]pyridines
Reprinted from: *Molecules* **2020**, *25*, 2194, doi:10.3390/molecules25092194 7

Yusuke Mukaijo, Soichi Yokoyama and Nagatoshi Nishiwaki
Comparison of Substituting Ability of Nitronate versus Enolate for Direct Substitution of a Nitro Group
Reprinted from: *Molecules* **2020**, *25*, 2048, doi:10.3390/molecules25092048 23

Eugene V. Babaev and Victor B. Rybakov
Phenacylation of 6-Methyl-Beta-Nitropyridin-2-Ones and Further Heterocyclization of Products
Reprinted from: *Molecules* **2020**, *25*, 1682, doi:10.3390/molecules25071682 33

Evgeny V. Pospelov, Ivan S. Golovanov, Sema L. Ioffe and Alexey Yu. Sukhorukov
The Cyclic Nitronate Route to Pharmaceutical Molecules: Synthesis of GSK's Potent PDE4 Inhibitor as a Case Study
Reprinted from: *Molecules* **2020**, *25*, 3613, doi:10.3390/molecules25163613 43

Yoshiki Sasaki, Masayoshi Takase, Shigeki Mori and Hidemitsu Uno
Synthesis and Properties of NitroHPHAC: The First Example of Substitution Reaction on HPHAC
Reprinted from: *Molecules* **2020**, *25*, 2486, doi:10.3390/molecules25112486 57

Lou Rocard, Antoine Goujon and Piétrick Hudhomme
Nitro-Perylenediimide: An Emerging Building Block for the Synthesis of Functional Organic Materials
Reprinted from: *Molecules* **2020**, *25*, 1402, doi:10.3390/molecules25061402 67

Feiyue Hao and Nagatoshi Nishiwaki
Recent Progress in Nitro-Promoted Direct Functionalization of Pyridones and Quinolones
Reprinted from: *Molecules* **2020**, *25*, 673, doi:10.3390/molecules25030673 85

About the Editor

Nagatoshi Nishiwaki, Professor, received his Ph.D. in 1991 from Osaka University. He worked at Professor Ariga's group in Department of Chemistry, Osaka Kyoiku University, as assistant professor (1991–2000) and as associate professor (2001–2008). From 2000 to 2001, he joined Karl Anker Jørgensen's group at Århus University in Denmark. He worked at Anan National College of Technology as associate professor from 2008 to 2009. Then, he moved to Kochi University of Technology in 2009, where he has been a professor since 2011. His research interests comprise synthetic organic chemistry using nitro compounds and heterocyclic compounds.

Editorial

A Walk through Recent Nitro Chemistry Advances

Nagatoshi Nishiwaki

Research Center for Molecular Design, Kochi University of Technology, Tosayamada, Kami, Kochi 782-8502, Japan; nishiwaki.nagatoshi@kochi-tech.ac.jp; Tel.: +81-887-57-2517

Received: 6 August 2020; Accepted: 7 August 2020; Published: 12 August 2020

Abstract: Chemistry of nitro groups and nitro compounds has long been intensively studied. Despite their long history, new reactions and methodologies are still being found today. This is due to the diverse reactivity of the nitro group. The importance of nitro chemistry will continue to increase in the future in terms of elaborate synthesis. In this article, we will take a walk through the recent advances in nitro chemistry that have been made in past decades.

Keywords: nitro group; conjugate addition; 1,3-Dipole; electron-withdrawing ability; electrophilicity; nitration; nitronate; nucleophilicity

1. Introduction

The chemistry of nitro compounds began at the beginning of the 19th century and has developed together with organic chemistry; in the 20th century, various reactivity properties of nitro groups were elucidated. Nitro compounds play an important role as building blocks and synthetic intermediates for the construction of scaffolds for drugs, agricultural chemicals, dyes, and explosives. In the world, millions of tons of nitro compounds are synthesized and consumed every year. In the 21st century, researchers' attentions gradually shifted to the use of nitro compounds in the elaborate syntheses such as controlling reactivity and stereochemistry. Development of new synthetic methods has also progressed using a combination of the diverse properties of nitro groups in past decades. Indeed, numerous methodologies are reported in current scientific journals. In this article, I would like to touch lightly on the recent advances in the chemistry of nitro compounds. For more information, please see the review articles cited in the references.

2. Nitration

Nitration is one of the fundamental chemical conversions. Conventional nitration processes involve HNO_3 alone or in combination with H_2SO_4, and this method has remained unchallenged for more than 150 years. Although other nitrating agents have been employed in a laboratory, these are not applicable to large-scale reactions because harsh conditions are sometimes necessary. The conventional methods also suffer from large amounts of waste acids and difficulty of regiocontrol [1,2]. These problems are overcome by using solid acids such as zeolites. High *para*-selective nitration was achieved by using tridirectional zeolites H_β [3] because of the steric restriction when substrate is adsorbed in the zeolite cavity [4].

Suzuki et al. developed an excellent nitration method using NO_2 and O_3, referred to as the Kyodai method [5]. This reaction proceeds efficiently even at low temperature. The addition of a small amount of a proton acid or Lewis acid enhances reactivity of the substrate to enable the polynitration.

Since nitrating agents also serve as strong oxidants, nitro compounds are often accompanied by oxidation products [6]. In order to avoid the formation of byproducts and regioisomers, *ipso*-nitration methods have been developed. Wu et al. showed metal-free nitration using phenylboronic acid and *t*-BuONO to afford nitrobenzene [7]. Furthermore, Buchwald et al. reported palladium catalyzed *ipso*-nitration method using chlorobenzene and commercially available $NaNO_2$ [8].

With the recent development of research on transition metal-catalyzed C-H activation, various skeletons have been constructed. Nitroaromatic compounds are obtained by this protocol, in which the directing group facilitates the regioselective nitration [9].

3. Reactivity and Application

The versatile reactivity of the nitro compounds family originates from the diverse properties of the nitro group. The strong electron-withdrawing nitro group reduces the electron density of the scaffold framework through both inductive and resonance effects, which undergoes reactions with nucleophiles or single-electron transfer. Makosza et al. indicated that reactions of nitroarenes with nucleophiles proceed through either direct nucleophilic attack forming σ-adduct or single-electron transfer forming a radical-ion pair [10–12].

The α-hydrogen is highly activated by the adjacent strong electron-withdrawing ability of the nitro group, which facilitates the α-arylation upon treatment of nitroalkanes with various arylating reagents leading to pharmaceutically active molecules [13]. The α-hydrogen is also acidic to attract basic reagents that are close together, and the spatial proximity undergoes an efficient reaction—similar to an intramolecular process—to afford polyfunctionalized compounds, which are referred to as a pseudo-intramolecular process [14]. The acidic hydrogen accelerates the tautomerism between nitroalkane and nitronic acid, among which the latter reveals high electrophilicity to react with carbon nucleophiles [15].

The nitro group stabilizes α-anion (nitronate ion), which serves as a nucleophile. Recently, the stereoselective Henry reaction (with aldehydes) [16,17] and nitro-Mannich reaction (with imine) [18] have been established, leading to enantiomerically rich β-nitroalcohols and β-nitroamines, respectively. Recent advances are noteworthy for the asymmetric organocatalytic conjugate addition of nitroalkanes to α,β-unsaturated carbonyl compounds [19,20]. Nitro group activates the connected carbon–carbon double bond, which serves as an excellent Michael acceptor to construct versatile frameworks [21–23]. These reactivities reveal significant utility in elaborate syntheses. Indeed, a lot of natural products have been synthesized using stereoselective reactions [24].

Nitro group also activates the connected carbon–carbon triple bond, however, it is too reactive to be used practically. The first synthesis of nitroalkyne was achieved in 1969 by Viehe [25]. During the subsequent half century, development of the synthetic methods and studies on reactivity, as well as physical/chemical properties, has progressed [26].

Deprotonated nitroalkane (nitronate) is characterized by the dual nature of nucleophilic and electrophilic properties. Indeed, versatile reactivities are used for synthesizing complex frameworks [27,28]. The dual nature of the nitronate also facilitates the 1,3-dipolar cycloadditions leading to functionalized heterocyclic compounds, which are not readily available by an alternative method [29,30].

Besides activating ability for the scaffold, the nitro group also serves as a good leaving group in organic reactions. A carbon–carbon double bond is formed upon the elimination of a HNO_2 from nitroalkane, which was energetically studied by Ballini et al. [12,31]. The combination of roles as an activator and as a leaving group enables the synthesis of polyfunctionalized compounds [32,33]. Furthermore, nitrobenzenes can be used as substrates for the transition-metal catalyzed cross-coupling, in which the nitro group is substituted with various nucleophiles [34].

Moreover, synthetic utility of the nitro group is improved by adding the chemical conversion to the abovementioned properties. The most fundamental transformation of the nitro group is reduction, which converts a nitro group to nitroso, oxime and amino groups. Vast numbers of combinations of catalysts and reducing agents have been developed for this purpose. Especially, recent progress of reduction using metal nanoparticles is noteworthy [35–37]. The landmark of the functional group conversion is the Nef reaction, which transforms a nitroalkane to the corresponding ketone. Since the first report at the end of 19th century [38], the usefulness of this reaction has not diminished, and it is still widely used in organic syntheses [39]. The chemical diversity of a nitro group enables us to construct a compound library possessing versatile electronic structure, which is helpful for developing new functional materials such as dyes and optical/electronic materials [40].

Due to the unique chemical behavior (reactivity and functional group conversions), nitro compounds serve as the synthetic intermediates for various types of compounds. In addition, nitro compounds themselves reveal specific properties. The explosive materials have been used in various situations such as the construction industry, mining minerals, processing metals and synthesis of nanomaterials, in which nitro compounds have played an important role [41]. Recent progress in this area provided more powerful explosive nitro compounds containing plural nitrogen. Although nitro compounds seem to be common in artificial materials, natural products containing a nitro group have been isolated from plants, fungi, bacteria, and mammals [42]. Accordingly, they exhibit biological activity. Indeed, many drugs containing a nitro group have been developed [43,44].

4. Conclusions

Chemistry of the nitro group and nitro compounds has been energetically investigated for a long time. Despite the long history including numerous reports, new reactions and methodologies are found even now. The unique physical/chemical properties of the nitro group will facilitate the progress of organic/inorganic chemistry and material science. Hence, nitro chemistry will continue to be increasingly important in the future.

Funding: This research received no external funding.

Conflicts of Interest: The author declares no conflict of interest.

References

1. Nishiwaki, N. Synthesis of Nitroso, Nitro, and Related Compounds. In *Comprehensive Organic Synthesis*, 2nd ed.; Molander, G.A., Knochel, P., Eds.; Elsevier: Oxford, UK, 2014; Volume 6, pp. 100–130.
2. Yan, G.; Yang, M. Recent Advances in the Synthesis of Aromatic Nitro Compounds. *Org. Biomol. Chem.* **2013**, *11*, 2554–2566. [CrossRef]
3. Smith, K.; Musson, A.; DeBoos, G.A. A Novel Method for the Nitration of Simple Aromatic Compounds. *J. Org. Chem.* **1998**, *63*, 8448–8454. [CrossRef]
4. Houas, M.; Kogelbauer, A.; Prins, R. An NMR Study of the Nitration of Toluene over Zeolites by HNO_3–Ac_2O. *Phys. Chem. Chem. Phys.* **2001**, *3*, 5067–5075. [CrossRef]
5. Shiri, M.; Zolfigol, M.A.; Kruger, H.G.; Tanbakouchian, Z. Advances in the Application of N_2O_4/NO_2 in Organic Reactions. *Tetrahedron* **2010**, *66*, 9077–9106. [CrossRef]
6. Prakash, G.K.S.; Mathew, T. ipso-Nitration of Arenes. *Angew. Chem. Int. Ed.* **2010**, *49*, 1726–1728. [CrossRef]
7. Wu, X.-F.; Schranck, J.; Neumann, H.; Beller, M. Convenient and Mild Synthesis of Nitroarenes by Metal-Free Nitration of Arylboronic Acids. *Chem. Commun.* **2011**, *47*, 12462–12463. [CrossRef] [PubMed]
8. Fors, B.P.; Buchwald, S.L. Pd-catalyzed Conversion of Aryl Chlorides, Triflates, and Nonaflates to Nitroaromatics. *J. Am. Chem. Soc.* **2009**, *131*, 12898–12899. [CrossRef] [PubMed]
9. Song, L.-R.; Fan, Z.; Zhang, A. Recent Advances in Transition Metal-Catalyzed $C(sp^2)$-H Nitration. *Org. Biomol. Chem.* **2019**, *17*, 1351–1361. [CrossRef] [PubMed]
10. Makosza, M. Reactions of Nucleophiles with Nitroarenes: Multifacial and Versatile Electrophiles. *Chem. Eur. J.* **2014**, *20*, 5536–5545. [CrossRef]
11. Makosza, M. How Does Nucleophilic Aromatic Substitution in Nitroarenes Really Proceed: General Mechanism. *Synthesis* **2017**, *49*, 3247–3254. [CrossRef]
12. Hao, F.; Nishiwaki, N. Recent Progress in Nitro-promoted Direct Functionalization of Pyridones and Quinolones. *Molecules* **2020**, *25*, 673. [CrossRef] [PubMed]
13. Zheng, P.-F.; An, Y.; Jiao, Z.-Y.; Shi, Z.-B.; Zhang, F.-M. Comprehension of the α-Arylation of Nitroalkanes. *Curr. Org. Chem.* **2019**, *23*, 1560–1580. [CrossRef]
14. Nishiwaki, N. Development of a Pseudo-Intramolecular Process. *Synthesis* **2016**, *48*, 1286–1300. [CrossRef]
15. Aksenov, N.A.; Aksenov, A.V.; Ovchanov, S.N.; Aksenov, D.A.; Rubin, M. Electrophilically Activated Nitroalkanes in Reactions with Carbon Based Nucleophiles. *Front. Chem.* **2020**, *8*, 77. [CrossRef] [PubMed]

16. Dong, L.; Chen, F.-E. Asymmetric Catalysis in Direct Nitromethane-Free Henry Reactions. *RSC Adv.* **2020**, *10*, 2313–2326. [CrossRef]
17. Ballini, R.; Gabrielli, S.; Palmieri, A.; Petrini, M. Nitroalkanes as Key Compounds for the Synthesis of Amino Derivatives. *Curr. Org. Chem.* **2011**, *15*, 1482–1506. [CrossRef]
18. Noble, A.; Anderson, J.C. Nitro-Mannich Reaction. *Chem. Rev.* **2013**, *113*, 2887–2939. [CrossRef]
19. Aitken, L.S.; Arezki, N.R.; Dell'Isola, A.; Cobb, A.J.A. Asymmetric Organocatalysis and the Nitro Group Functionality. *Synthesis* **2013**, *45*, 2627–2648.
20. Roca-Lopez, D.; Sadaba, D.; Delso, I.; Herrera, R.P.; Tejero, T.; Merino, P. Asymmetric organocatalytic synthesis of γ-nitrocarbonyl compounds through Michael and Domino reactions. *Tetrahedron Asymmetry* **2010**, *21*, 2561–2601. [CrossRef]
21. Halimehjani, A.Z.; Namboothiri, I.N.N.; Hooshmand, S.E. Nitroalkenes in the synthesis of carbocyclic compounds. *RSC Adv.* **2014**, *4*, 31261–31299. [CrossRef]
22. Ballini, R.; Araújo, N.; Gil, M.V.; Román, E.; Serrano, J.A. Conjugated nitrodienes. Synthesis and reactivity. *Chem. Rev.* **2013**, *113*, 3493–3515. [CrossRef] [PubMed]
23. Nakaike, Y.; Asahara, H.; Nishiwaki, N. Construction of Push-Pull Systems Using β-Formyl-β-nitroenamine. *Russ. Chem. Bull. Int. Ed.* **2016**, *65*, 2129–2142. [CrossRef]
24. Sukhorukov, A.Y.; Sukhanova, A.A.; Zlotin, S.G. Stereoselective Reactions of Nitro Compounds in the Synthesis of Natural Compound Analogs and Active Pharmaceutical Ingredients. *Tetrahedron* **2016**, *72*, 6191–6281. [CrossRef]
25. Jaeger, V.; Viehe, H.G. Heterosubstituted Acetylenes. XXI. Nitroacetylenes. *Angew. Chem. Int. Ed.* **1969**, *8*, 273–274.
26. Windler, G.K.; Pagoria, P.F.; Vollhardt, K.P.C. Nitroalkynes: A Unique Class of Energetic Materials. *Synthesis* **2014**, *46*, 2383–2412. [CrossRef]
27. Tabolin, A.A.; Sukhorukov, A.Y.; Ioffe, S.L. α-Electrophilic Reactivity of Nitronates. *Chem. Rec.* **2018**, *18*, 1489–1500. [CrossRef]
28. Sukhorukov, A.Y. C-H reactivity of the α-Position in Nitrones and Nitronates. *Adv. Synth. Catal.* **2020**, *362*, 724–754. [CrossRef]
29. Tabolin, A.A.; Sukhorukov, A.Y.; Ioffe, S.L.; Dilman, A.D. Recent Advances in the Synthesis and Chemistry of Nitronates. *Synthesis* **2017**, *49*, 3255–3268. [CrossRef]
30. Baiazitov, R.Y.; Denmark, S.E. Tandem [4+2]/[3+2] Cycloadditions. In *Methods and Applications of Cycloaddition Reactions in Organic Syntheses*; Nishiwaki, N., Ed.; John Wiley & Sons: Hoboken, NJ, USA, 2014; pp. 471–550.
31. Ballini, R.; Palmieri, A. Formation of Carbon-Carbon Double Bonds: Recent Developments *via* Nitrous Acid Elimination (NAE) from Aliphatic Nitro Compounds. *Adv. Synth. Catal.* **2019**, *361*, 5070–5097. [CrossRef]
32. Mukaijo, Y.; Yokoyama, S.; Nishiwaki, N. Comparison of Substituting Ability of Nitronate versus Enolate for Direct Substitution of a Nitro Group. *Molecules* **2020**, *25*, 2048. [CrossRef]
33. Asahara, H.; Sofue, A.; Kuroda, Y.; Nishiwaki, N. Alkynylation and Cyanation of Alkenes Using Diverse Properties of a Nitro Group. *J. Org. Chem.* **2018**, *83*, 13691–13699. [CrossRef] [PubMed]
34. Yang, Y. Palladium-Catalyzed Cross-Coupling of Nitroarenes. *Angew. Chem. Int. Ed.* **2017**, *56*, 15802–15804. [CrossRef] [PubMed]
35. Formenti, D.; Ferretti, F.; Scharnagl, F.K.; Beller, M. Reduction of nitro compounds using 3d-non-noble metal catalysts. *Chem. Rev.* **2019**, *119*, 2611–2680. [CrossRef] [PubMed]
36. Orlandi, M.; Brenna, D.; Harms, R.; Jost, S.; Benaglia, M. Recent Developments in the Reduction of Aromatic and Aliphatic Nitro Comppounds to Amines. *Org. Process Res. Dev.* **2018**, *22*, 430–445. [CrossRef]
37. Aditya, T.; Pal, A.; Pal, T. Nitroarene Reduction: A Trusted Model Reaction to Test Nanoparticle Catalysts. *Chem. Commun.* **2015**, *51*, 9410–9431. [CrossRef] [PubMed]
38. Nef, J.U. Ueber die Constitution der Salze der Nitroparaffine. *Justus Liebigs Ann. Chem.* **1894**, *280*, 263–291. [CrossRef]
39. Ballini, R.; Petrini, M. The Nitro to Carbonyl Conversion (Nef Reaction): New Perspectives for a Classical Transformation. *Adv. Synth. Catal.* **2015**, *357*, 2371–2402. [CrossRef]
40. Rocar, L.; Goujon, A.; Hudhomme, P. Nitro-Perylenediimide: An Emerging Buiding Block for the Synthesis of Functional Organic Materials. *Molecules* **2020**, *25*, 1402. [CrossRef]
41. Kumar, D.; Elias, A.J. The Explosive Chemistry of Nitrogen. *Resonance* **2019**, 1253–1271. [CrossRef]

42. Parry, R.; Nishino, S.; Spain, J. Naturally-Occurring Nitro Compounds. *Nat. Prod. Rep.* **2011**, *28*, 152–167. [CrossRef]
43. Nepali, K.; Lee, H.-Y.; Liou, J.-P. Nitro-Group-Containing Drugs. *J. Med. Chem.* **2019**, *62*, 2851–2893. [PubMed]
44. Patterson, S.; Wyllie, S. Nitro Drugs for the Treatment of Trypanosomatid Diseases: Past, Present, and Future Prospects. *Trends Parasitol.* **2014**, *30*, 289–298. [CrossRef] [PubMed]

© 2020 by the author. Licensee MDPI, Basel, Switzerland. This article is an open access article distributed under the terms and conditions of the Creative Commons Attribution (CC BY) license (http://creativecommons.org/licenses/by/4.0/).

Article

Synthesis and Facile Dearomatization of Highly Electrophilic Nitroisoxazolo[4,3-*b*]pyridines

Maxim A. Bastrakov [1,*], Alexey K. Fedorenko [1,2], Alexey M. Starosotnikov [1], Ivan V. Fedyanin [3] and Vladimir A. Kokorekin [1]

1. N.D. Zelinsky Institute of Organic Chemistry RAS; Leninsky prosp. 47, 119991 Moscow, Russia; alexeyfedorenko21@mail.ru (A.K.F.); alexey41@list.ru (A.M.S.); kokorekin@yandex.ru (V.A.K.)
2. Chemistry Department, Lomonosov Moscow State University, Leninskie gory, 1/3, 119991 Moscow, Russia
3. A.N. Nesmeyanov Institute of Organoelement Compounds, Vavilova str. 28, 119991 Moscow, Russia; octy@xrlab.ineos.ac.ru
* Correspondence: b_max82@mail.ru; Tel.: +7-(499)-135-5328

Received: 22 April 2020; Accepted: 7 May 2020; Published: 8 May 2020

Abstract: A number of novel 6-R-isoxazolo[4,3-*b*]pyridines were synthesized and their reactions with neutral C-nucleophiles (1,3-dicarbonyl compounds, π-excessive (het)arenes, dienes) were studied. The reaction rate was found to be dependent on the nature of the substituent 6-R. The most reactive 6-nitroisoxazolo[4,3-*b*]pyridines are able to add C-nucleophiles in the absence of a base under mild conditions. In addition, these compounds readily undergo [4+2]-cycloaddition reactions on aromatic bonds C=C(NO$_2$) of the pyridine ring, thus indicating the superelectrophilic nature of 6-NO$_2$-isoxazolo[4,3-*b*]pyridines.

Keywords: nitro group; nitropyridines; isoxazolo[4,3-*b*]pyridines; 1,4-dihydropyridines; nucleophilic addition; Diels-Alder reaction; dearomatization

1. Introduction

The nitro group is considered to be a versatile and unique functional group in organic chemistry. Synthetic and natural compounds containing nitro groups display great structural diversity [1,2], and they exhibit a wide range of biological activities [3] including antibiotic [1], antitumor [1,4], and anti-HIV activities [5–7]. In addition, nitroarenes are used as agrochemical preparations [8,9], energetic compounds [10] and in the production of innovative materials [11].

It is well known that the introduction of one or more nitro groups in aromatic or heteroaromatic nucleus increases the electron-deficient character of the molecule. Such compounds have been extensively studied in recent decades due to their interesting, sometimes exceptional, properties. Their high susceptibility to undergoing nucleophilic addition or substitution processes with very weak nucleophiles has raised considerable interest, leading to numerous synthetic, biological, and analytical applications [12–31].

Such compounds possess extremely high reactivity towards carbon and heteroatomic nucleophiles, therefore a special term, "superelectrophile", was coined in order to distinguish them from other electrophilic aromatics [26,32]. Typical examples of such compounds are given below (Figure 1).

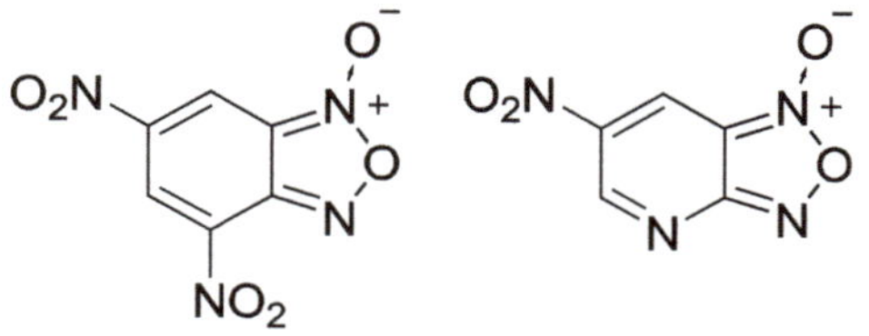

Figure 1. Selected examples of superelectrophiles.

In addition, these compounds are capable to undergo [4+2]-cycloadditions to the C=C(NO$_2$) aromatic bond, behaving as electron-poor dienophiles with dienes, or as heterodienes with electron-rich dienophiles within normal or inverse electronic demands, respectively [16,33–35]. The above-mentioned interactions with nucleophiles or dienes resulted in dearomatization of the initial aromatic nitro compound. At the same time, dearomatization as a method of converting accessible, cheap, and simple aromatic compounds into more saturated, inaccessible and promising intermediates of greater molecular complexity is a very important approach in modern organic chemistry [36,37].

This work is part of our ongoing research on highly electrophilic systems and the application of the dearomatization strategy in the synthesis of new polyfunctional azaheterocycles [38–48]. We have previously shown that nitropyridines fused with π-deficient heterocycles (furoxan **A**, selenadiazole **B**), Scheme 1, react with neutral nucleophiles with the formation of 1,4-addition products—dihydropyridine derivatives [45,46,48].

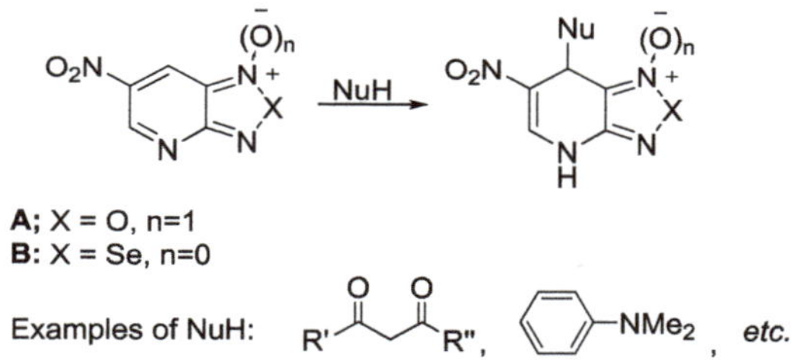

A; X = O, n=1
B: X = Se, n=0

Examples of NuH: R'C(O)CH$_2$C(O)R'', C$_6$H$_5$-NMe$_2$, etc.

Scheme 1. Reactions of condensed nitropyridines with nucleophiles.

Another possible condensed pyridines structurally close to heterocyclic systems **A** and **B** and presumably having a similar electron-deficient character are isoxazolo[4,3-b]pyridines **C**, Figure 2. The present work is devoted to the synthesis of pyridine derivatives condensed with an isoxazole ring and study of their interaction with various neutral C-nucleophiles as well as their behavior in [4+2]-cycloaddition reactions.

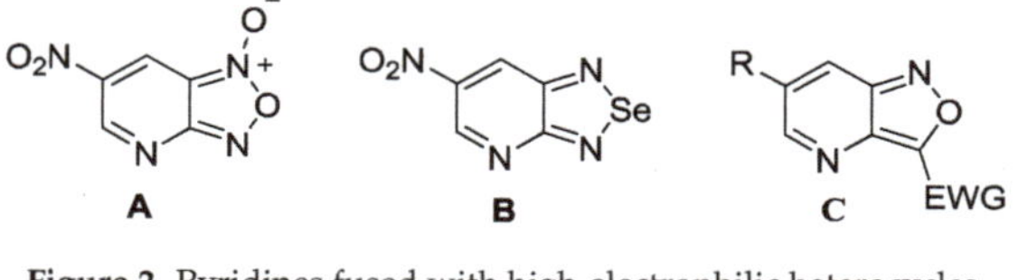

Figure 2. Pyridines fused with high-electrophilic heterocycles.

2. Results and Discussion

2.1. Synthesis of 6-R-Isoxazolo[4,3-b]pyridines 3a–j

6-R-Isoxazolo[4,3-b]pyridines **3a–j** were synthesized according to a two-steps procedure, previously described in the literature for **3j** [49]. Commercially available 2-chloro-3-nitropyridines **1a–e** used as starting compounds were involved in Sonogashira cross-coupling with terminal alkynes to give 2-alkynylpyridines **2a–j**. In turn, the cycloisomerization of compounds **2a–j** in the presence of catalytic amounts of iodine(I) chloride gave the desired 6-R-3-acylisoxazolo[4,3-b]pyridines **3a–j** in good yields, Scheme 2, Table 1.

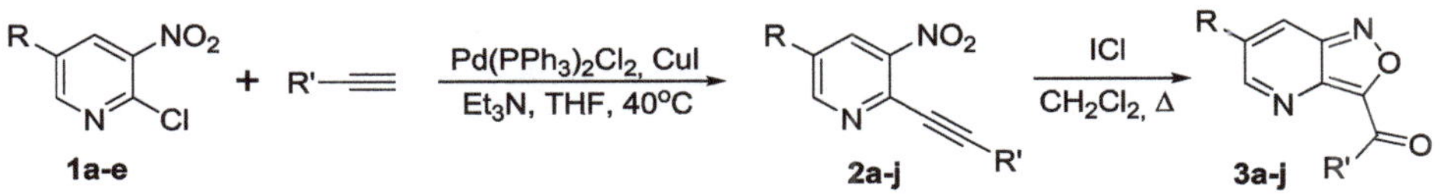

Scheme 2. Synthesis of 6-R-isoxazolo[4,3-b]pyridines **3a–j**.

Table 1. Isolated yields of compounds **2a–j** and **3a–j**.

Compound 1	R	R'	Product 2, Yield (%)	Product 3, Yield (%)
1a	NO_2	Ph	2a, 72	3a, 85
1a	NO_2	4-Me-C_6H_4	2b, 63	3b, 87
1a	NO_2	4-F-C_6H_4	2c, 61	3c, 72
1a	NO_2	c-C_3H_5	2d, 84	3d, 74
1a	NO_2	c-C_5H_9	2e, 82	3e, 71 *
1a	NO_2	n-C_5H_{11}	2f, 35	3f, 80 *
1b	CO_2Me	Ph	2g, 76	3g, 65
1c	CF_3	Ph	2h, 42	3h, 73
1d	Cl	Ph	2i, 40	3i, 60
1e	H	Ph	2j, 82	3j, 80

* The yield is shown for the crude product.

In the case of compounds **2e** and **2f**, ^{1}H NMR spectroscopy showed that, along with the expected isoxazolo[4,3-b]pyridines, the formation of minor unidentified products (5–10%) occurred. All attempts to isolate target compounds in their pure forms failed, therefore compounds **3e,f** were used without further purification. The structure of compounds **2** and **3** was established on the basis of NMR and HRMS data, and for compounds **2a**, **2c**, **3b** it was additionally confirmed by X-Ray analysis.

2.2. X-ray of *2a, 2c, 3b*

The crystals of **2a** and **2c** are isostructural with minor differences in the unit cell parameters. All bonds, bond angles and torsion angles are typical as confirmed by a Mogul geometry check [50]. The bond angles at the triple bond C2-C7-C8 (176.11(14) and 176.37(15)°) and C7-C8-C9 (174.20(14) and 171.90(16)° in **2a** and **2c**) deviate from the idealized value of 180° for linear conformation. The angles between the average planes of nitro groups and pyridine group are within the range 7.09(17)–12.44(14)°, despite the presence of the short intramolecular contact O1···C7 (2.6763(17) and 2.6580(18) Å in **2a** and **2c**). Pyridine and phenyl rings are nearly co-planar with interplane angles equal to 5.97(5) and 5.61(5)°. In crystal packing, a head-to-tail arrangement of the molecules is observed with π-stacking interaction between formally acceptor dinitro substituted pyridine ring and phenyl moieties (C···C from ca. 3.3 Å). All other intermolecular contacts are weak and non-directional.

The crystal of **3b** is a first example of determined crystal structure containing isoxazolo[4,3-b]pyridine ring, Figure 3. The distribution of bond distances in the heterocycle (N1-O2 1.4109(18), O2-C3 1.3487(19), C3-C3A 1.378(2), N1-C7A 1.330(2), C3A-C7A 1.426(2) Å) is quite similar to the one found in a number of benzo[c]isoxazoles found in Cambridge Structural Database and confirms the canonical structure shown in Figure 3. Due to steric reasons, the heterocycle and tolyl substituents are non-coplanar with O2-C3-C8-C9 torsion angle equal to 48.4(2)°. In crystal molecules, infinite π-stacks (C···C from ca. 3.4 Å) of alternating molecules with head-to-tail arrangement of heterocycles and tolyl fragments are formed.

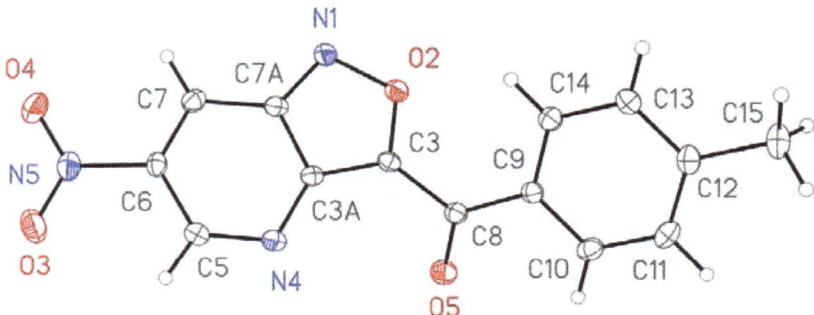

Figure 3. General view of **3b** in crystal. Anisotropic displacement parameters for non-hydrogen atoms are drawn at 50% probability.

2.3. Nucleophilic Addition to 6-R-Isoxaxolo[4,3-b]pyridines

We studied the interaction of isoxazolo[4,3-*b*]pyridines **3** with neutral C-nucleophiles: CH acids and π-excessive (het)arenes. It was found that nitro derivatives **3a–d** react with all ranges of used nucleophiles under mild conditions (MeCN, room temperature, base-free), forming 1,4-addition products, **4a–m**, Scheme 3. As in the case of **A** [45], β-dicarbonyl compounds react with **3** in enolic form. The reaction rate was similar to that of superelectrophiles **A** and **B** [45,46] (Scheme 1), thus indicating the high electrophilicity of isoxazolo[4,3-*b*]pyridine system. Some of the reactions proceeded almost immediately after mixing the reagents, the others were completed within an hour.

The methoxycarbonyl derivative **3g** forms adducts **4n,o** with 1,3-dicarbonyl compounds somewhat slower: full conversion of starting material required 2–3 h without the addition of a base. 6-Unsubstituted isoxazolopyridine **3j** gave the adduct **4p** with most acidic dimedone after 4 h stirring, Scheme 3. Surprisingly, we were unable to isolate any adducts of isoxazolo[4,3-*b*]pyridines **3h** and **3i** containing electron-withdrawing Cl and CF₃ groups in position 6. The application of more drastic conditions (MeCN, 80 °C) was not effective; the starting compounds were recovered. The reason for the observed reactivity is not clear, however, we can conclude that the ability of 6-R-isoxazolo[4,3-*b*]pyridines to add neutral C-nucleophiles depends on the substituent 6-R and decreases in the following order

$$NO_2 > CO_2Me > H \gg Cl, CF_3$$

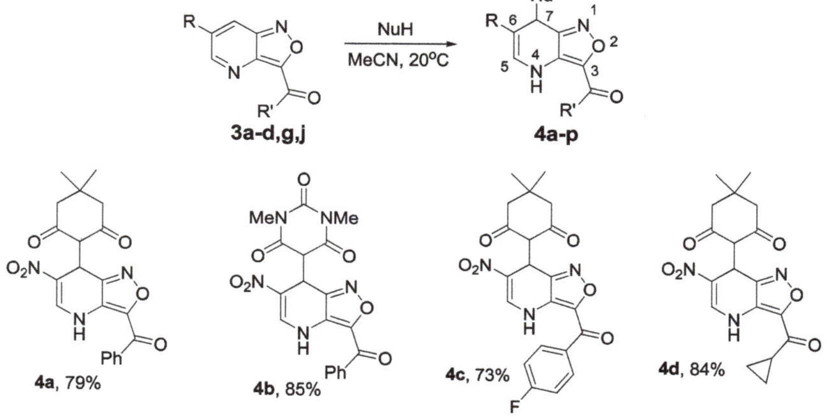

Scheme 3. *Cont.*

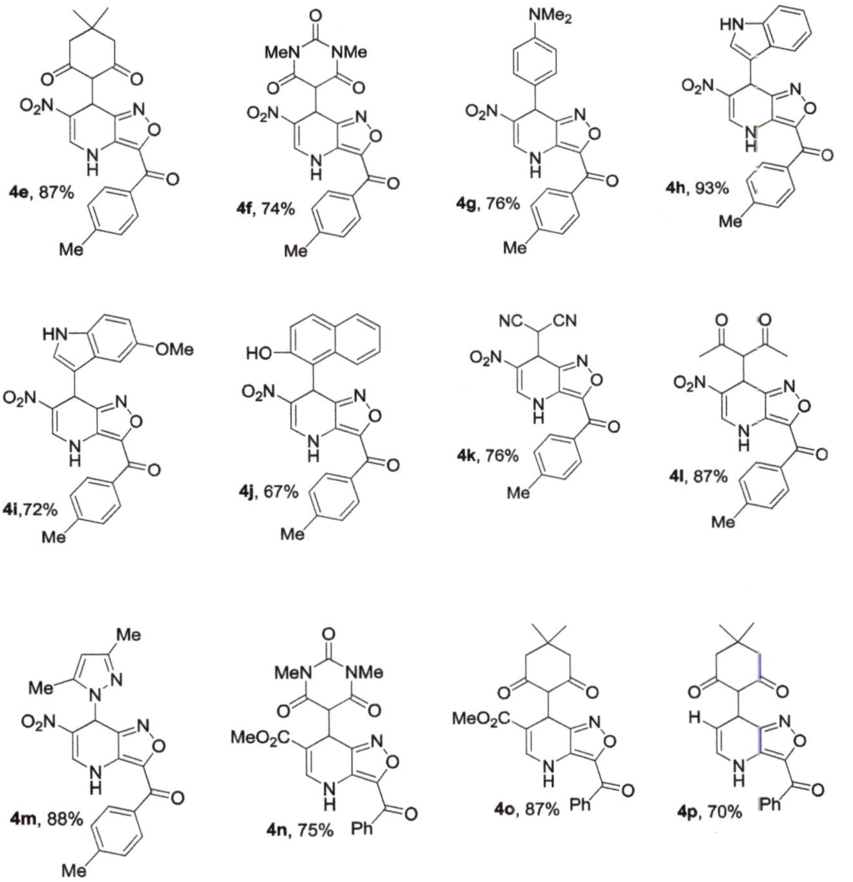

Scheme 3. Reactions of 6-R-isoxazolo[4,3-b]pyridines **3** with nucleophiles.

The structures of compounds **4** were established on the basis of NMR spectroscopy and HRMS data. In ^{1}H NMR spectra of adducts **4**, the signals corresponding to H(7) protons in the range of 5.0–5.5 ppm, as well as downfield signals of NH protons (9.8–10.4 ppm) and H(5) at 8.1 ppm, were observed as doublets with close coupling constants. This confirms the nucleophilic addition at position 7 and is consistent with the results obtained previously for other highly electrophilic azolopyridines [45,46,48].

2.4. 6-NO$_2$-Isoxazolo[4,3-b]pyridines in Diels-Alder Reactions

The ability to add weak (neutral) nucleophiles is one of the features inherent to superelectrophilic aromatic systems. However, increasing the electrophilicity leads to a decrease in aromaticity. Therefore, (hetero)aromatic superelectrophiles are prone to undergo [4+2]-cycloaddition with dienes or nucleophilic dienophiles [16,26,32–35].

We found that only 6-NO$_2$-isoxazolo[4,3-b]pyridines **3a–f** are able to give cycloaddition products in Diels-Alder reactions with 2,3-dimethyl-1,3-butadiene while compounds **3g–j** with other substituents at position 6 were unreactive. This once again highlights the originality of the nitro group among other electron-withdrawing functional groups and its impact on the electrophilicity of the aromatic systems.

Reactions of compounds **3a–f** with 2,3-dimethyl-1,3-butadiene were carried out in CH$_2$Cl$_2$ or CHCl$_3$ at room temperature (Scheme 4, Table 2). The C=C–NO$_2$ fragment of a pyridine ring acts as a dienophile and the process proceeds in accordance with normal electron demands. However, in all cases, instead of the expected adducts **5**, we isolated compounds **6a–g** —products of the further addition of H_2O to a C=N-double bond.

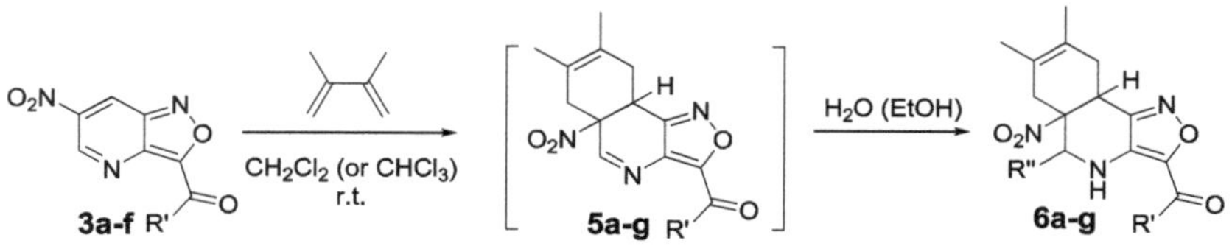

Scheme 4. [4+2]-Cycloaddition reactions of 6-NO_2-isoxazolo[4,3-b]pyridines **3a–f**.

Table 2. Isolated yields of compounds **6a–g**.

Compound 3	R'	R"	Product 6, Yield (%)
3a	Ph	OH	6a, 74
3b	4-Me-C_6H_4	OH	6b, 80
3c	4-F-C_6H_4	OH	6c, 73
3d	c-C_3H_5	OH	6d, 80
3e	c-C_5H_9	OH	6e, 51
3f	n-C_5H_{11}	OH	6f, 35
3b	4-Me-C_6H_4	OEt	6g, 42

The intermediate adducts **5a–g** are likely to be unstable and exhibit an extremely high tendency to react with nucleophiles (e.g., water) to form compounds **6a–g** in good yields. Carrying out the reaction in an inert atmosphere, the additional purification of all solvents did not allow us to isolate compounds **5**. Apparently the formation of products **6** occurs on contact with air moisture at the isolation step. Reaction of **3b** with dimethylbutadiene in chloroform (stabilized with 1.5% EtOH) gave compound **6g**—the product of EtOH addition. In our opinion, this fact indirectly confirms the hypothesis of the high electrophilicity of compounds **5**.

The structure of cycloadducts **6** was proved by 2D NMR spectroscopy experiments (COSY, ^{1}H-^{13}C HMBC, ^{1}H-^{13}C HSQC). For compounds **6a** and **6d**, the full assignment of hydrogen and carbon atoms in the NMR spectra was made. NMR experiments confirmed the proposed addition of a diene at the C=C(NO_2) bond, Figure 4.

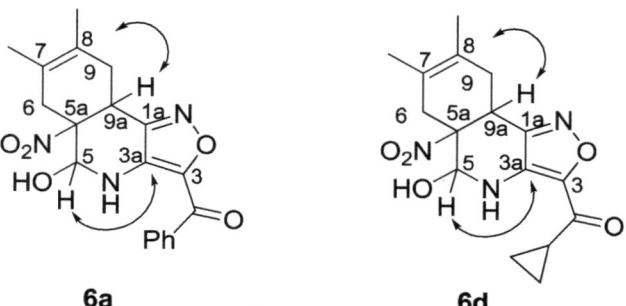

Figure 4. Selected interactions in 2D HMBC spectra of compounds **6a,d**.

Cross peaks corresponding to H(5)-C(3a), as well as H(9a)-C(1a) and H(9a)-C(8) interactions were observed in the ^{1}H-^{13}C HMBC spectra of these compounds. In addition, we observed the coupling of two nonequivalent protons H(9) with the H(9a) proton. Such interactions were described earlier for similar cycloadducts of pyridofuroxan **A** (Figure 2) [34]. These data allowed us to make an unambiguous conclusion about the direction of cycloaddition and hydration.

3. Conclusions

A number of new 6-R-isoxazolo[4,3-*b*]pyridines were synthesized, starting from 2-chloro-3-nitropyridines. It was found that, in reactions with neutral C-nucleophiles, the reactivity of 6-R-isoxazolo[4,3-*b*]pyridines depends on the nature of the substituent 6-R. 6-Nitro derivatives were found to add 1,3-dicarbonyl compounds and π-excessive arenes and hetarenes to the pyridine ring under mild conditions to form the corresponding 1,4-adducts. In addition, reactions with 2,3-dimethyl-1,3-butadiene led to [2+4]-adducts on the aromatic bonds C=C(NO$_2$) of the pyridine ring. The condensed 3,4-dihydropyridines thus formed easily, adding the molecule of water to C=N double bond to give polyfunctionalized tetrahydropyridine derivatives. All the above properties of 6-nitroisoxazolo[4,3-*b*]pyridines make it possible to classify compounds of this class as superelectrophiles.

4. Materials and Methods

4.1. General Information

All chemicals were of commercial grade and used directly without purification. Melting points were measured on a Stuart SMP 20 apparatus. ^{1}H and ^{13}C NMR spectra were recorded on Bruker AC-200 (at 200 and 50 MHz, respectively), Bruker AM-300 (at 300.13 and 75.13 MHz, respectively), Bruker Avance DRX 500 (at 500 and 125 MHz, respectively) or Bruker Avance II 600 spectrometer (at 600 and 150 MHz, respectively) in DMSO-d$_6$ or CDCl$_3$. IR spectra were recorded on BrukerAlpha spectrometer, and the samples were prepared as KBr pellets. HRMS spectra were recorded on a Bruker micrOTOF II mass spectrometer using ESI. All reactions were monitored by TLC analysis using ALUGRAM SIL G/UV254 plates, which were visualized by UV light. Compounds **1a–e** were purchased from commercial suppliers. Compounds **2j** and **3j** were synthesized according to the method [48]. X-ray data collection was performed on a Bruker APEX II diffractometer equipped with Apex II CCD detector and operating with MoKα radiation (λ = 0.71073 Å). Frames were integrated using the Bruker SAINT software package [51] by a narrow-frame algorithm. A semi-empirical absorption correction was applied with the SADABS program [52] using the intensity data of equivalent reflections. The structures were solved with a dual-space approach with SHELXT program [53] and refined by the full-matrix least-squares technique against F^2_{hkl} in anisotropic approximation with SHELXL [54] software package. All hydrogen atoms were placed in calculated positions and refined in the riding model, with U_{iso}(H) constrained to be 1.5U_{eq} and 1.2U_{eq} of the parent methyl and all other carbon atoms. Detailed crystallographic information is given in Table S3 in Supplementary Materials. Crystallographic data have been deposited to the Cambridge Crystallographic Data Centre, CCDC 1983530-1983532 can be retrieved free of charge via https://www.ccdc.cam.ac.uk/structures.

4.2. Synthesis of Compounds **2a–i**

A mixture of the appropriate 2-chloropyridine **1** (5 mmol), PdCl$_2$(PPh$_3$)$_2$ (0.17 g; 5 mol-%), and Et$_3$N (1.01 g; 10 mmol) was suspended in anhydrous THF (20 mL). The appropriate acetylene (5.5 mmol) was then injected under argon, followed by addition of CuI (0.02 g; 2.5 mol-%). The reaction mixture was stirred under argon at 40 °C temperature until full completion (1–3 h, completion observed by TLC). Solvent was evaporated under the reduced pressure; the crude residue was purified by column chromatography (elution by chloroform).

3,5-Dinitro-2-(phenylethynyl)pyridine (**2a**). 72%. Orange powder. M.p. 183–185 °C. ^{1}H NMR (300 MHz, CDCl$_3$): δ 7.45–7.54 (m, 3H, Ph.), 7.74–7.76 (d, *J* = 7.2 Hz, 2H, Ph), 9.17 (s, 1H, H4), 9.63 (s, 1H, H6). ^{13}C NMR (75 MHz, DMSO-d$_6$): δ 85.1, 100.9, 119.9, 128.7, 129.2, 131.3, 132.4, 142.3, 146.2, 148.3. HRMS (ESI) calc. for [C$_{13}$H$_8$N$_3$O$_4$]$^+$ [M + H]$^+$ 270.0509, found 270.0514.

3,5-Dinitro-2-(p-tolylethynyl)pyridine (**2b**). 63%. Orange powder. M.p. 203–205 °C. ^{1}H NMR (300 MHz, CDCl$_3$): δ 2.46 (s, 3H, Me), 7.30 (d, J = 8.0 Hz, 2H, p-Tolyl), 7.66 (d, J = 8.0 Hz, 2H, p-Tolyl), 9.18 (d, J = 2.2 Hz, 1H, H4), 9.63 (d, J = 2.2 Hz, 1H, H6). ^{13}C NMR (75 MHz, DMSO-d$_6$): δ 21.3, 84.9, 101.6, 116.8, 128.6, 129.8, 132.4, 139.8, 141.7, 142.1, 146.0, 148.3. HRMS (ESI) calc. for [C$_{14}$H$_{10}$N$_3$O$_4$]$^+$ [M + H]$^+$ 284.0666, found 284.0669.

3,5-Dinitro-2-((4-fluorophenyl)ethynyl)pyridine (**2c**). 61%. Orange powder. M.p. 170–172 °C. ^{1}H NMR (200 MHz, DMSO-d$_6$): δ 7.40 (t, J = 8.5 Hz, 2H, 4F-Ph), 7.78 (dd, J = 7.9, 5.7 Hz, 2H, 4F-Ph), 9.21 (d, J = 2.2 Hz, 1H, H4), 9.65 (d, J = 2.2 Hz, 1H, H6). ^{13}C NMR (75 MHz, DMSO-d$_6$): δ 85.0, 99.8, 128.8, 116.7 (d, $^2J_{C-F}$ = 22.5 Hz), 116.6, 116.0, 130.6, 135.2 (d, $^3J_{C-F}$ = 9.2 Hz), 139.6, 142.3, 146.1, 147.2, 148.4, 163.5 (d, $^1J_{C-F}$ = 251.5 Hz). HRMS (ESI) calc. for [C$_{13}$H$_7$FN$_3$O$_4$]$^+$ [M + H]$^+$ 288.0415, found 288.0417.

2-(Cyclopropylethynyl)-3,5-dinitropyridine (**2d**). 84%. Orange powder. M.p. 128–130 °C. ^{1}H NMR (300 MHz, CDCl$_3$): δ 1.09–1.18 (m, 4H), 1.68 (dt, J = 13.1, 6.6 Hz, 1H), 9.08 (d, J = 2.1 Hz, 1H, H4), 9.53 (d, J = 2.1 Hz, 1H, H6). ^{13}C NMR (75 MHz, DMSO-d$_6$): δ 0.4, 9.9, 72.1, 109.5, 128.4, 141.7, 148.0. HRMS (ESI) calc. for [C$_{10}$H$_8$N$_3$O$_4$]$^+$ [M + H]$^+$ 234.0509, found 234.0517.

2-(Cyclopentylethynyl)-3,5-dinitropyridine (**2e**). 32%. Orange oil. ^{1}H NMR (300 MHz, CDCl$_3$): δ 1.75–1.64 (m, 2H), 1.87 (m, 4H), 2.10 (m, 2H), 3.04 (p, J = 7.2 Hz, 1H), 9.08 (d, J = 2.1 Hz, 1H, H4), 9.55 (d, J = 2.1 Hz, 1H, H6). ^{13}C NMR (75 MHz, CDCl$_3$): δ 25.3, 31.2, 33.2, 76.5, 112.5, 127.9, 141.3, 142.0, 146.1, 147.7. HRMS (ESI) calc. for [C$_{12}$H$_{12}$N$_3$O$_4$]$^+$ [M + H]$^+$ 262.0822, found 262.0816.

2-(Hept-1-yn-1-yl)-3,5-dinitropyridine (**2f**). 35%. Orange oil. ^{1}H NMR (300 MHz, CDCl$_3$): δ 0.78–1.06 (m, 3H), 1.25–1.55 (m, 4H), 1.74 (p, J = 7.0 Hz, 2H), 2.63 (t, J = 7.1 Hz, 2H), 9.08 (d, J = 2.1 Hz, 1H, H4), 9.56 (d, J = 2.1 Hz, 1H, H6). ^{13}C NMR (75 MHz, CDCl$_3$): δ 14.0, 20.4, 22.3, 27.5, 31.2, 108.8, 112.6, 128.0, 128.4, 141.5, 142.2, 147.8 HRMS (ESI) calc. for [C$_{12}$H$_{14}$N$_3$O$_4$]$^+$ [M + H]$^+$ 264.0979, found 264.0970.

Methyl-5-nitro-6-(phenylethynyl)nicotinate (**2g**) 76%. Orange powder. M.p. 117–119 °C. ^{1}H NMR (300 MHz, CDCl$_3$): δ 4.05 (s, 3H, Me), 7.42–7.52 (m, 3H, Ph.), 7.71–7.74 (d, J = 7.2 Hz, 2H, Ph), 8.95 (d, J = 1,3 Hz, 1H, H4), 9.39 (d, J = 1.3 Hz, 1H, H6). ^{13}C NMR (75 MHz, CDCl$_3$): δ 53.2, 85.4, 101.4, 121.0, 125.0, 128.7, 130.7, 133.0, 133.5, 140.5, 146.5, 153.9, 163.5. HRMS (ESI) calc. for [C$_{15}$H$_{11}$N$_2$O$_4$]$^+$ [M + H]$^+$ 283.0713, found 283.0721.

3-Nitro-2-(phenylethynyl)-5-(trifluoromethyl)pyridine (**2h**) 42% Yellow powder. M.p. 124–126 °C. ^{1}H NMR (300 MHz, CDCl$_3$): δ 7.47 (m, 3H, Ph), 7.73 (d, J = 6.5 Hz, 2H, Ph), 8.65 (s, 1H, H4), 9.09 (s, 1H, H6). ^{13}C NMR (126 MHz, CDCl$_3$): δ 84.8, 101.5, 120.9, 124.4 (q, $^1J_{C-F}$ = 273.0 Hz), 125.8, 128.8, 130.4 (q, $^3J_{C-F}$ = 3.7 Hz), 131.0, 133.1, 140.6, 146.1, 149.9 (q, $^3J_{C-F}$ = 3.5 Hz). HRMS (ESI) calc. for [C$_{14}$H$_8$F$_3$N$_2$O$_2$]$^+$ [M + H]$^+$ 293.0532, found 293.0542.

5-Chloro-3-nitro-2-(phenylethynyl)pyridine (**2i**) 40%. M.p. 103–105 °C. ^{1}H NMR (300 MHz, CDCl$_3$): δ 7.41–7.48 (m, 3H, Ph.), 7.69–7.71 (d, J = 6.8 Hz, 2H, Ph), 8.41 (d, J = 1.6 Hz, 1H, H4), 8.81 (d, J = 1.6 Hz, 1H, H6). ^{13}C NMR (75 MHz, CDCl$_3$): δ 84.5, 99.3, 121.3, 128.7, 130.4, 130.9, 132.3, 132.8, 135.5, 152.7. HRMS (ESI) calc. for [C$_{13}$H$_8$ClN$_2$O$_2$]$^+$ [M + H]$^+$ 259.0269, found 259.0259.

4.3. Synthesis of Compounds 3a–i

Iodine monochloride (19.5 mg, 0.12 mmol) was added to a solution of the appropriate compound **2** (4 mmol) in dichloromethane (20 mL), and the resulting solution was heated under reflux until full completion (4–8 h.). Solvent was evaporated under the reduced pressure; the crude residue was purified by column chromatography (elution by dichloromethane).

(6-Nitroisoxazolo[4,3-b]pyridin-3-yl)(phenyl)methanone (**3a**) 85% Yellowish powder. M.p. 135–137 °C. ^{1}H NMR (300 MHz, CDCl$_3$): δ 7.63 (t, J = 7.7 Hz, 2H, Ph.), 7.77 (t, J = 7.4 Hz, 1H, Ph), 8.23 (d, J = 7.4 Hz, 2H, Ph) 9.08 (d, J = 2.2 Hz, 1H, H5), 9.55 (d, J = 2.2 Hz, 1H, H7). ^{13}C NMR (75 MHz, CDCl$_3$): δ 122.7, 129.7, 131.2, 134.9, 135.7, 135.9, 144.7, 149.6, 150.4, 163.0, 181.1. HRMS (ESI) calc. for [C$_{13}$H$_8$N$_3$O$_4$]$^+$ [M + H]$^+$ 270.0509, found 270.0508.

(6-Nitroisoxazolo[4,3-b]pyridin-3-yl)(p-tolyl)methanone (**3b**) 87% Yellowish powder. M.p. 158–160 °C. ^{1}H NMR (300 MHz, CDCl$_3$): δ 2.52 (s, 3H, Me), 7.43 (d, J = 8.1 Hz, 2H, p-Tolyl), 8.14 (d, J = 8.2 Hz, 2H, p-Tolyl), 9.07 (d, J = 2.2 Hz, 1H, H5), 9.53 (d, J = 2.2 Hz, 1H, H7). ^{13}C NMR (151 MHz, CDCl$_3$): δ 22.1, 122.1, 129.9, 130.8, 132.8, 134.1, 144.0, 146.5, 148.8, 149.7, 162.8, 180.0. HRMS (ESI) calc. for [C$_{14}$H$_{10}$N$_3$O$_4$]$^+$ [M + H]$^+$ 284.0666, found 284.0669.

(4-Fluorophenyl)(6-nitroisoxazolo[4,3-b]pyridin-3-yl)methanone (**3c**) 72%. Yellowish powder. M.p. 115–117 °C. ^{1}H NMR (300 MHz, CDCl$_3$): δ 7.33 (d, J = 8.5 Hz, 2H, 4F-Ph), 8.32 (dd, J = 8.8, 5.3 Hz, 2H, 4F-Ph), 9.09 (d, J = 2.2 Hz, 1H, H5), 9.55 (d, J = 2.2 Hz, 1H, H7). ^{13}C NMR (75 MHz, DMSO-d$_6$): δ 116.5 (d, $^2J_{C-F}$ = 22.2 Hz), 122.1, 131.8, 133.5 (d, $^3J_{C-F}$ = 9.8 Hz), 134.3, 144.2, 149.1, 149.8, 163.8 (d, $^1J_{C-F}$ = 225.4 Hz), 168.7, 178.8. HRMS (ESI) calc. for [C$_{13}$H$_7$FN$_3$O$_4$]$^+$ [M + H]$^+$ 288.0415, found 288.0411.

Cyclopropyl(6-nitroisoxazolo[4,3-b]pyridin-3-yl)methanone (**3d**) 74%. Yellowish powder. M.p. 118–120 °C. ^{1}H NMR (300 MHz, CDCl$_3$): δ 1.32-1.37 (m, 2H), 1.53–1.56 (m, 2H), 3.43–3.51 (m, 1H), 9.06 (d, J = 2.2 Hz, 1H, H5), 9.54 (d, J = 2.2 Hz, 1H, H7). ^{13}C NMR (75 MHz, CDCl$_3$): δ 14.3, 20.3, 122.3, 133.4, 148.9, 150.0, 161.4, 188.0. HRMS (ESI) calc. for [C$_{10}$H$_8$N$_3$O$_4$]$^+$ [M + H]$^+$ 234.0509, found 234.0521.

Methyl-3-benzoylisoxazolo[4,3-b]pyridine-6-carboxylate (**3g**) 65%. Yellowish powder. M.p. 114–116 °C. ^{1}H NMR (300 MHz, CDCl$_3$): δ 4.07 (s, 3H, Me), 7.61 (t, J = 7.6 Hz, 2H, Ph), 7.74 (t, J = 7.4 Hz, 1H, Ph), 8.25 (d, J = 7.5 Hz, 2H, Ph), 8.85 (d, J = 1.7 Hz, 1H, H5), 9.36 (d, J = 1.7 Hz, 1H, H7). ^{13}C NMR (75 MHz, CDCl$_3$): δ 53.3, 127.6, 127.8, 128.9, 130.6, 134.6, 135.7, 150.8, 155.0, 161.4, 164.2, 165.8, 180.8. HRMS (ESI) calc. for [C$_{15}$H$_{11}$N$_2$O$_4$]$^+$ [M + H]$^+$ 283.0713, found 283.0709.

Phenyl(6-(trifluoromethyl)isoxazolo[4,3-b]pyridin-3-yl)methanone (**3h**) 73%. Yellowish powder. M.p. 117–119 °C. ^{1}H NMR (300 MHz, CDCl$_3$): δ 7.62 (t, J = 7.6 Hz, 2H, Ph), 7.75 (t, J = 7.3 Hz, 1H, Ph), 8.24 (d, J = 7.8 Hz, 2H, Ph), 8.52 (s, 1H, H5), 8.98 (s, 1H, H7). ^{13}C NMR (75 MHz, CDCl$_3$): δ 122.5 (q, $^1J_{C-F}$ = 273.6 Hz) 123.5 (q, $^2J_{C-F}$ = 3.1 Hz), 126.1, 128.4, 128.7, 129.1, 130.0, 130.6, 134.1, 134.8, 135.6, 149.7, 151.1, 151.1, 161.9, 180.8. HRMS (ESI) calc. for [C$_{14}$H$_8$F$_3$N$_2$O$_2$]$^+$ [M + H]$^+$ 293.0532, found 293.0533.

(6-Chloroisoxazolo[4,3-b]pyridin-3-yl)(phenyl)methanone (**3i**) 60%. Beige powder. M.p. 120–122 °C. ^{1}H NMR (300 MHz, CDCl$_3$): δ 7.61 (t, J = 7.6 Hz, 2H, Ph), 7.74 (t, J = 7.4 Hz, 1H, Ph), 8.14 (d, J = 1.7 Hz, 1H, H5), 8.23 (d, J = 7.5 Hz, 2H, Ph), 8.71 (s, 1H, H7). ^{13}C NMR (75 MHz, CDCl$_3$): δ 121.8, 129.0, 130.6, 131.9, 133.8, 134.7, 135.7, 151.5, 155.5, 161.1, 180.9. HRMS (ESI) calc. for [C$_{13}$H$_8$ClN$_2$O$_2$]$^+$ [M + H]$^+$ 259.0269, found 259.0276.

4.4. Synthesis of Compounds 4a–p

A mixture of the appropriate isoxazole **3** (0.5 mmol) and nucleophile (0.5 mmol) was dissolved in anhydrous CH$_3$CN (5 mL). The reaction mixture was stirred at r.t. until full completion (1–3 h, by TLC). The solution was diluted with water (25 mL), and the obtained precipitate was filtered off.

2-(3-Benzoyl-6-nitro-4,7-dihydroisoxazolo[4,3-b]pyridin-7-yl)-5,5-dimethylcyclohexane-1,3-dione (**4a**) 79%. Yellow powder. M.p. 244–246 °C. ^{1}H NMR (DMSO-d$_6$): δ 0.93 (s, 6H, 2Me), 2.25 (br.s, 4H, 2CH$_2$), 5.75 (br.s, 1H, H7), 7.64 (t, J = 7.5 Hz, 2H, Ph), 7.75 (t, J = 7.3 Hz, 1H, Ph), 8.02 (d, J = 5.7 Hz, 1H, H5), 8.14 (d, J = 7.7 Hz, 2H, Ph), 10.56 (d, J = 6.1 Hz, NH). ^{13}C NMR (75 MHz, DMSO-d$_6$): δ 27.3, 28.0, 31.8, 42.6, 49.8, 50.0, 126.8, 129.1, 129.3, 134.2, 135.2, 137.5, 137.6, 146.3, 157.5, 172.6, 173.1, 180.8. HRMS (ESI) calc. for [C$_{21}$H$_{20}$N$_3$O$_6$]$^+$ [M + H]$^+$ 410.1347, found 410.1340.

5-(3-Benzoyl-6-nitro-4,7-dihydroisoxazolo[4,3-b]pyridin-7-yl)-1,3-dimethylpyrimidine-2,4,6-(1H,3H,5H)-trione (**4b**) 85%. Yellow powder. M.p. 234–235 °C. ^{1}H NMR (300 MHz, DMSO-d$_6$): δ 3.03 (s, 3H, Me), 3.17 (s, 3H, Me), 4.65 (s, 1H, CH), 5.52 (s, 1H, H7), 7.64 (t, J = 7.2 Hz, 2H, Ph), 7.76 (t, J = 7.2 Hz, 1H, Ph), 8.13 (d, J = 7.1 Hz, 3H, Ph and H5), 10.93 (d, J = 3.7 Hz, 1H, NH). ^{13}C NMR (75 MHz, DMSO-d$_6$): δ 28.1, 28.3, 33.5, 33.8, 33.9, 52.4, 129.1, 129.4, 134.4, 139.7, 151.1, 166.2, 166.6, 180.6. HRMS (ESI) calc. for [C$_{19}$H$_{16}$N$_5$O$_7$]$^+$ [M + H]$^+$ 426.1044, found 426.1037.

2-(3-(4-Fluorobenzoyl)-6-nitro-4,7-dihydroisoxazolo[4,3-b]pyridin-7-yl)-5,5-dimethylcyclohexane-1,3-dione (**4c**) 73%. Yellow powder. M.p. 251–253 °C. ^{1}H NMR (300 MHz, DMSO-d$_6$): δ 0.93 (s, 6H, 2Me), 2.24 (br.s, 4H, 2CH$_2$), 5.75 (s, 1H, H7), 7.48 (t, *J* = 8.8 Hz, 2H, 4F-Ph), 8.01 (s, 1H, H5), 8.23 (dd, *J* = 8.7, 5.6 Hz, 2H, 4F-Ph), 10.54 (d, *J* = 1.5 Hz, 1H, NH). ^{13}C NMR (75 MHz, DMSO-d$_6$): δ 27.4, 27.5, 31.7, 116.2 (d, $^2J_{C–F}$ = 22.1 Hz), 126.6, 126.8, 131.8, 132.4 (d, $^3J_{C–F}$ = 9.8 Hz), 137.3, 146.1, 157.4, 165.4 (d, $^1J_{C–F}$ = 253.5 Hz) 179.1. HRMS (ESI) calc. for [C$_{21}$H$_{19}$FN$_3$O$_6$]$^+$ [M + H]$^+$ 428.1252, found 428.1256.

2-(3-(Cyclopropanecarbonyl)-6-nitro-4,7-dihydroisoxazolo[4,3-b]pyridin-7-yl)-5,5-dimethylcyclohexane-1,3-dione (**4d**) 84%. Yellow powder. M.p. 258–260 °C. ^{1}H NMR (300 MHz, DMSO-d$_6$): δ 0.92 (s, 6H, 2Me), 1.16 (s, 4H, 2CH$_2$ (c-Pr)), 2.26 (br.s, 4H, 2CH$_2$), 2.65 (s, 1H, CH), 5.71 (br.s, 1H, H7), 7.95 (s, 1H, H5), 10.39 (br.s, 1H, NH). ^{13}C NMR (75 MHz, DMSO-d$_6$): δ 12.0, 12.3, 18.2, 27.6, 31.7, 122.8, 126.1, 137.5, 146.2, 157.65, 189.6. HRMS (ESI) calc. for [C$_{18}$H$_{20}$N$_3$O$_6$]$^+$ [M + H]$^+$ 374.1346, found 374.1341.

5,5-Dimethyl-2-(3-(4-methylbenzoyl)-6-nitro-4,7-dihydroisoxazolo[4,3-b]pyridin-7-yl)-cyclohexane-1,3-dione (**4e**) 87%. Yellow powder. M.p. 225–227 °C. ^{1}H NMR (300 MHz, DMSO-d$_6$): δ 0.93 (s, 6H, 2Me), 2.43 (s, 3H, Me, *p*-Tolyl), 5.75 (s, 1H, H7), 7.45 (d, *J* = 7.2 Hz, 2H, *p*-Tolyl), 7.98–8.13 (m, 3H, *p*-Tolyl and H5), 10.49 (d, *J* = 6.7 Hz, 1H, NH). ^{13}C NMR (75 MHz, DMSO-d$_6$): δ 21.3, 27.2, 31.7, 129.4, 129.6, 132.6, 144.9, 157.3, 180.2. HRMS (ESI) calc. for [C$_{22}$H$_{22}$N$_3$O$_6$]$^+$ [M + H]$^+$ 424.1503, found 424.1505.

1,3-Dimethyl-5-(3-(4-methylbenzoyl)-6-nitro-4,7-dihydroisoxazolo[4,3-b]pyridin-7-yl)-pyrimidine-2,4,6(1H,3H,5H)-trione (**4f**) 74%. Yellow powder. M.p. 241–243 °C. ^{1}H NMR (300 MHz, DMSO-d$_6$): δ 2.43 (s, 3H, Me, *p*-Tolyl), 3.03 (s, 3H, Me), 3.17 (s, 3H, Me), 4.64 (s, 1H, CH), 5.51 (s, 1H, H7), 7.45 (d, *J* = 8.1 Hz, 2H, *p*-Tolyl), 8.06 (d, *J* = 8.1 Hz, 2H, *p*-Tolyl), 8.14 (d, *J* = 6.4 Hz, 1H, H5), 10.90 (d, *J* = 6.5 Hz, 1H, NH). ^{13}C NMR (75 MHz, DMSO-d$_6$): δ 21.3, 28.1, 28.3, 33.8, 52.4, 124.1, 125.4, 129.6, 129.7, 132.3, 139.7, 145.3, 148.1, 151.1, 155.1, 166.2, 166.6, 180.1. HRMS (ESI) calc. for [C$_{20}$H$_{18}$N$_5$O$_7$]$^+$ [M + H]$^+$ 440.1201, found 440.1197.

(7-(4-(Dimethylamino)phenyl)-6-nitro-4,7-dihydroisoxazolo[4,3-b]pyridin-3-yl)(p-tolyl)-methanone (**4g**) 76%. Orange powder. M.p. 237–239 °C. ^{1}H NMR (200 MHz, DMSO-d$_6$): δ 2.40 (s, 3H, Me, p-Tolyl), 2.83 (s, 6H, 2Me), 5.62 (s, 1H, H7), 6.64 (d, *J* = 8.7 Hz, 2H, Ar), 7.08 (d, *J* = 8.4 Hz, 2H, *p*-Tolyl), 7.42 (d, *J* = 8.1 Hz, 2H, Ar), 8.04 (d, *J* = 8.1 Hz, 2H, *p*-Tolyl), 8.21 (s, 1H. H5), 10.68 (s, 1H, NH). ^{13}C NMR (75 MHz, DMSO-d$_6$): δ 21.3, 37.6, 40.1, 112.5, 124.4, 127.2, 127.7, 129.1, 129.5, 129.6, 132.4, 137.2, 145.0, 148.1, 149.7, 156.8, 180.1. HRMS (ESI) calc. for [C$_{22}$H$_{21}$N$_4$O$_4$]$^+$ [M + H]$^+$ 405.1557, found 405.1564.

(7-(1H-indol-3-yl)-6-nitro-4,7-dihydroisoxazolo[4,3-b]pyridin-3-yl)(p-tolyl)methanone (**4h**) 93%. Yellow powder. M.p. 221–223 °C. ^{1}H NMR (300 MHz, DMSO-d$_6$): δ 2.40 (s, 3H, Me), 6.03 (s, 1H, H7), 6.99–7.38 (m, 7H, indole and *p*-Tolyl), 8.05 (s, 2H, *p*-Tolyl), 8.23 (s, 1H, H5), 10.75 (s, 1H, NH), 11.07 (s, 1H, NH). ^{13}C NMR (126 MHz, DMSO-d$_6$): δ 21.3, 30.4, 66.3, 111.8, 114.2, 118.1, 119.0, 121.3, 123.4, 123.6, 124.7, 125.0, 126.7, 129.5, 129.6, 132.4, 136.4, 136.9, 145.0, 147.9, 156.4, 180.1. HRMS (ESI) calc. for [C$_{22}$H$_{17}$N$_4$O$_4$]$^+$ [M + H]$^+$ 401.1243, found 401.1240.

(7-(5-Methoxy-1H-indol-3-yl)-6-nitro-4,7-dihydroisoxazolo[4,3-b]pyridin-3-yl)(p-tolyl)-methanone (**4i**) 72%. Yellow powder. M.p. 181–183 °C. ^{1}H NMR (300 MHz, DMSO-d$_6$): δ 2.41 (s, 3H, Me), 3.71 (s, 3H, OMe), 6.00 (s, 1H, H7), 6.74 (d, *J* = 9.3 Hz, 1H, indole H6), 6.92 (s, 1H, indole H4), 7.24 (m, 2H, indole H2 and H7), 7.43 (d, *J* = 7.5 Hz, 2H, *p*-Tolyl), 8.06 (d, *J* = 7.8 Hz, 2H, *p*-Tolyl), 8.23 (s, 1H, H5), 10.78 (s, 1H, NH), 10.91 (s, 1H, NH). ^{13}C NMR (75 MHz, DMSO-d$_6$): δ 21.3, 30.4, 55.2, 66.3, 100.2, 111.1, 112.5, 114.1, 124.1, 124.8, 125.4, 126.7, 129.6, 129.7, 131.6, 132.5, 136.9, 145.1, 153.3, 156.5, 180.2. HRMS (ESI) calc. for [C$_{23}$H$_{19}$N$_4$O$_5$]$^+$ [M + H]$^+$ 431.1349, found 431.1340.

(7-(2-Hydroxynaphthalen-1-yl)-6-nitro-4,7-dihydroisoxazolo[4,3-b]pyridin-3-yl)(p-tolyl)-methanone (**4j**) 67%. Yellow powder. M.p. 204–206 °C. ^{1}H NMR (300 MHz, DMSO-d$_6$): δ 2.41 (s, 3H, Me), 6.62 (s, 1H, H7), 7.00 (d, *J* = 9.2 Hz, 1H, Ar H3), 7.35 (t, *J* = 7.5 Hz, 1H, Ar H6), 7.44 (d, *J* = 8.0 Hz, 2H, *p*-Tolyl), 7.58 (t, *J* = 7.8 Hz, 1H, Ar H5), 7.72 (d, *J* = 8.8 Hz, 1H, Ar H8), 7.81 (d, *J* = 8.2 Hz, 1H, Ar), 8.06 (d, *J* = 8.2 Hz, 2H, *p*-Tolyl), 8.17 (s, 1H, H7), 8.53 (d, *J* = 8.9 Hz, 1H, Ar H4), 9.89 (s, NH). ^{13}C NMR (75 MHz,

DMSO-d$_6$): δ 20.2, 29.4, 117.2, 117.7, 121.4, 121.7, 125.6, 125.8, 126.7, 127.0, 128.2, 128.3, 128.4, 128.5, 131.4, 131.8, 136.3, 136.5, 143.9, 145.2, 151.9, 152.1, 156.3, 179.1. HRMS (ESI) calc. for [C$_{24}$H$_{18}$N$_3$O$_5$]$^+$ [M + H]$^+$ 428.1240, found 428.1237.

2-(3-(4-Methylbenzoyl)-6-nitro-4,7-dihydroisoxazolo[4,3-b]pyridin-7-yl)malononitrile (**4k**) 76%. Yellow powder. M.p. 214–216 °C. ^{1}H NMR (300 MHz, DMSO-d$_6$): δ 2.45 (s, 3H, Me), 5.59 (s, 1H, CH), 5.67 (s, 1H, H7), 7.49 (d, *J* = 6.1 Hz, 2H, *p*-Tolyl), 8.12 (d, *J* = 6.7 Hz, 2H, *p*-Tolyl), 8.32 (s, 1H, H5), 11.23 (s, 1H, NH). ^{13}C NMR (75 MHz, DMSO-d$_6$): δ 21.4, 27.9, 35.1, 112.1, 112.2, 120.5, 124.7, 129.7, 132.1, 140.8, 145.5, 149.5, 151.6, 180.0. Found, %: C, 58.39; H, 3.22; N, 20.07; C$_{17}$H$_{11}$N$_5$O$_4$ Calc., %: C, 58.46; H, 3.17; N, 20.05.

3-(3-(4-Methylbenzoyl)-6-nitro-4,7-dihydroisoxazolo[4,3-b]pyridin-7-yl)pentane-2,4-dione (**4l**) 87%. Yellow powder. M.p. 188–190 °C. ^{1}H NMR (300 MHz, DMSO-d$_6$): δ 2.05 (s, 3H, Me, *p*-Tolyl), 2.42 (d, *J* = 6.4 Hz, 6H, 2Me), 4.90 (s, 1H, CH), 5.30 (s, 1H, H7), 7.45 (d, *J* = 8.1 Hz, 2H, *p*-Tolyl), 8.07 (d, *J* = 7.1 Hz, 3H, *p*-Tolyl and H5), 10.80 (s, 1H, NH). ^{13}C NMR (75 MHz, DMSO-d$_6$): δ 21.3, 29.6, 31.1, 32.0, 66.9, 124.1, 126.3, 129.6, 132.3, 139.6, 145.2, 154.9, 180.1, 203.2, 204.8. HRMS (ESI) calc. for [C$_{19}$H$_{17}$N$_3$O$_6$+NH$_4$]$^+$ [M + NH$_4$]$^+$ 401.1453, found 401.1456.

(7-(3,5-Dimethyl-1H-pyrazol-1-yl)-6-nitro-4,7-dihydroisoxazolo[4,3-b]pyridin-3-yl)(p-tolyl)-methanone (**4m**) 88%. Yellow powder. M.p. 202–204 °C. ^{1}H NMR (300 MHz, DMSO-d$_6$): δ 1.95 (s, 3H, Me), 2.05 (s, 3H, Me, *p*-Tolyl), 2.42 (s, 3H, Me), 5.82 (s, 1H, H7), 7.14 (s, 1H), 7.44 (d, *J* = 8.1 Hz, 2H, *p*-Tolyl), 8.07 (d, *J* = 8.1 Hz, 2H, *p*-Tolyl), 8.34 (s, 1H) 11.19 (s, 1H, NH). Found, %: C, 58.25; H, 4.22; N, 19.07; C$_{18}$H$_{15}$N$_5$O$_4$ Calc., %: C, 59.18; H, 4.14; N, 19.17.

Methyl-3-benzoyl-7-(1,3-dimethyl-2,4,6-trioxohexahydropyrimidin-5-yl)-4,7-dihydroisoxazolo[4,3-b]pyridine-6-carboxylate (**4n**) 75%. Beige powder. M.p. 170–172 °C. ^{1}H NMR (300 MHz, DMSO-d$_6$): δ 3.03 (s, 3H, Me), 3.13 (s, 3H, Me), 3.63 (s, 3H, CO$_2$Me), 4.43 (s, 1H, CH), 5.07 (s, 1H, H7), 7.44 (d, *J* = 5.4 Hz, 1H, H5), 7.63 (t, *J* = 7.3 Hz, 2H, Ph), 7.74 (t, *J* = 7.0 Hz, 1H, Ph), 8.12 (d, *J* = 7.7 Hz, 2H, Ph), 10.2 (d, *J* = 5.3 Hz, 1H, NH). ^{13}C NMR (75 MHz, DMSO-d$_6$): δ 28.0, 28.1, 33.3, 51.3, 54.0, 99.1, 127.2, 129.0, 129.2, 134.0, 135.2, 138.9, 151.4, 154.9, 166.4, 167.8, 180.5. HRMS (ESI) calc. for [C$_{21}$H$_{19}$N$_4$O$_7$]$^+$ [M + H]$^+$ 439.1248, found 439.1242.

Methyl-3-benzoyl-7-(4,4-dimethyl-2,6-dioxocyclohexyl)-4,7-dihydroisoxazolo[4,3-b]pyridine-6-carboxylate (**4o**) 87%. Beige powder. M.p. 232–234 °C. ^{1}H NMR (300 MHz, DMSO-d$_6$): δ 0.93 (s, 6H, 2Me), 2.16 (br.s, 4H, 2CH$_2$), 3.53 (s, 3H, CO$_2$Me), 5.36 (s, 1H, H7), 7.33 (s, 1H, H5), 7.61 (t, *J* = 7.3 Hz, 2H, Ph), 7.70 (t, *J* = 7.1 Hz, 1H, Ph), 8.11 (d, *J* = 7.6 Hz, 2H, Ph), 9.73 (d, *J* = 5.0 Hz, NH). ^{13}C NMR (75 MHz, DMSO-d$_6$): δ 26.1, 27.3, 28.3, 31.7, 40.0, 50.7, 102.6, 128.7, 128.8, 129.0, 133.7, 135.6, 136.6, 136.7, 144.5, 157.4, 166.3, 180.4. HRMS (ESI) calc. for [C$_{23}$H$_{23}$N$_2$O$_6$]$^+$ [M + H]$^+$ 423.1551, found 423.1545.

2-(3-Benzoyl-4,7-dihydroisoxazolo[4,3-b]pyridin-7-yl)-5,5-dimethylcyclohexane-1,3-dione (**4p**) 70%. Beige powder. M.p. 180–182 °C. ^{1}H NMR (300 MHz, DMSO-d$_6$): δ 0.91 (s, 3H, Me), 0.99 (s, 3H, Me), 2.08–2.40 (m, 4H, 2CH$_2$), 4.33 (s, 1H, CH), 5.76 (s, 1H, H7), 7.56-7.67 (m, 4H, Ph and H6), 8.08 (d, *J* = 6.4 Hz, 2H, Ph), 8.44 (s, 1H, H5). ^{13}C NMR (75 MHz, DMSO-d$_6$): δ 19.4, 24.6, 27.0, 28.5, 31.9, 41.3, 49.7, 77.0, 110.4, 124.2, 126.8, 128.7, 128.8, 128.9, 130.3, 133.2, 134.4, 156.5, 169.5, 180.0, 194.2. HRMS (ESI) calc. for [C$_{21}$H$_{21}$N$_2$O$_4$]$^+$ [M + H]$^+$ 365.1496, found 365.1494.

4.5. Synthesis of Compounds 6a–g

2,3-Dimethylbutadiene (0.5 mL, 4.5 mmol) was added to a solution of the appropriate isoxazolopyridine **3** (0.5 mmol) in dichloromethane (or CHCl$_3$) (5 mL). The reaction mixture was stirred at r.t. until full completion (normally 4–8 h, TLC control). The solution was diluted with hexane (15 mL), and the obtained precipitate was filtered off.

(5-Hydroxy-7,8-dimethyl-5a-nitro-4,5,5a,6,9,9a-hexahydroisoxazolo[4,3-c]isoquinolin-3-yl)-(phenyl)methanone (**6a**) 74%. Beige powder. M.p. 173–175 °C. ^{1}H NMR (300 MHz, CDCl$_3$): δ 1.56 (s, 3H, Me), 1.67 (s, 3H, Me), 2.45 (d, *J* = 18.1 Hz, 1H), 2.77–3.06 (m, 3H), 3.27 (d, *J* = 3.9 Hz, 1H, OH), 4.35 (d, *J* = 6.8 Hz, 1H), 5.41 (t,

J = 4.0 Hz, 1H), 6.68 (d, J = 3.3 Hz, 1H), 7.55 (t, J = 7.6 Hz, 2H, Ph), 7.65 (t, J = 7.3 Hz, 1H, Ph), 8.27 (d, J = 7.2 Hz, 2H, Ph). ^{13}C NMR (75 MHz, CDCl3): δ 18.8, 19.0, 29.3, 30.2, 35.1, 78.6, 89.3, 119.6, 124.2, 128.9, 129.8, 131.5, 133.9, 135.6, 146.7, 154.3, 181.8. HRMS (ESI) calc. for [$C_{19}H_{20}N_3O_5$] + [M + H]$^+$ 370.1397, found 370.1397.

(5-Hydroxy-7,8-dimethyl-5a-nitro-4,5,5a,6,9,9a-hexahydroisoxazolo[4,3-c]isoquinolin-3-yl)-(p-tolyl)methanone (**6b**) 80%. Beige powder. M.p. 148–150 °C. ^{1}H NMR (300 MHz, CDCl3): δ 1.56 (s, 3H, Me), 1.66 (s, 3H, Me), 2.47–3.06 (m, 7H, 2CH$_2$+Me(*p*-Tolyl)), 3.25 (d, J = 3.9 Hz, 1H, OH), 4.34 (d, J = 6.9 Hz, 1H), 5.39 (t, J = 3.6 Hz, 1H), 6.65 (d, J = 4.0 Hz, 1H, NH), 7.35 (d, J = 8.2 Hz, 2H, *p*-Tolyl), 8.20 (d, J = 8.1 Hz, 2H, *p*-Tolyl). ^{13}C NMR (75 MHz, CDCl3): δ 18.8, 19.0, 21.9, 29.3, 30.2, 35.1, 78.7, 89.4, 119.6, 124.2, 129.6, 130.0, 131.3, 133.1, 144.9, 146.8, 154.3, 181.4. HRMS (ESI) calc. for [$C_{20}H_{22}N_3O_5$]$^+$ [M + H]$^+$ 384.1553, found 384.1548.

(4-Fluorophenyl)(5-hydroxy-7,8-dimethyl-5a-nitro-4,5,5a,6,9,9a-hexahydroisoxazolo[4,3-c]-isoquinolin-3-yl) methanone (**6c**) 73%. Beige powder. M.p. 177–179 °C. ^{1}H NMR (300 MHz, CDCl3): δ 1.56 (s, 3H, Me), 1.67 (s, 3H, Me), 2.47 (d, J = 17.4Hz, 1H), 2.78–2.91 (m, 3H, 2CH$_2$), 3.18 (d, J = 3.6 Hz, 1H, OH), 4.34 (d, J = 6.9 Hz, 1H), 5.41 (t, J = 3.3 Hz, 1H), 6.65 (d, J = 3.2 Hz, 1H, NH), 7.23 (t, J = 8.6 Hz, 2H, 4F-Ph), 8.34 (dd, J = 8.7, 5.4 Hz, 2H, 4F-Ph). ^{13}C NMR (75 MHz, CDCl3): δ 18.8, 19.0, 29.4, 30.2, 35.1, 78.6, 89.3, 116.2 (d, $^2J_{C-F}$ = 21.9 Hz), 119.6, 124.2, 131.5, 131.9, 132.0, 132.6 (d, $^3J_{C-F}$ = 9.5 Hz), 146.5, 154.3, 166.2 (d, $^1J_{C-F}$ = 256.5 Hz) 180.0. HRMS (ESI) calc. for [$C_{19}H_{19}FN_3O_5$]$^+$ [M + H]$^+$ 388.1303, found 388.1316.

Cyclopropyl-(5-hydroxy-7,8-dimethyl-5a-nitro-4,5,5a,6,9,9a-hexahydroisoxazolo[4,3-c]-isoquinolin-3-yl)methanone (**6d**) 80%. Beige powder. M.p. 148–150 °C. ^{1}H NMR (300 MHz, CDCl3): δ 1.12–1.27 (m, 4H, c-Pr), 1.55 (s, 3H, Me), 1.65 (s, 3H, Me), 2.43 (d, J = 18.6 Hz, 1H), 2.61-3.00 (m, 4H, 2CH$_2$+1H(c-Pr)), 3.67 (s, 1H, OH), 4.30 (d, J = 6.9 Hz, 1H), 5.31 (d, J = 3.5 Hz, 1H), 6.19 (d, J = 3.4 Hz, 1H, NH). ^{13}C NMR (75 MHz, CDCl3): δ 12.2, 12.4, 18.2, 18.8, 18.9, 29.3, 30.1, 35.1, 78.6, 89.3, 119.5, 124.2, 127.4, 146.8, 154.7, 191.5. HRMS (ESI) calc. for [$C_{16}H_{20}N_3O_5$+NH$_4$]$^+$ [M + NH$_4$]$^+$ 351.1663, found 351.1668.

Cyclopentyl(5-hydroxy-7,8-dimethyl-5a-nitro-4,5,5a,6,9,9a-hexahydroisoxazolo[4,3-c]-isoquinolin-3-yl)methanone (**6e**) 51%. Beige powder. M.p. 105–107 °C. 1H NMR (300 MHz, CDCl3): δ 1.55 (s, 3H, Me), 1.60–1.73 (m, 7H, Me+2CH$_2$), 1.80–2.08 (m, 4H, 2CH$_2$), 2.43 (d, J = 18.0 Hz, 1H), 2.76-3.04 (m, 3H, 2CH$_2$), 3.39 (s, 1H, OH), 3.47-3.59 (m, 1H, CH), 4.29 (d, J = 7.1 Hz, 1H), 5.34 (d, J = 3.3 Hz, 1H), 6.23 (d, J = 3.3 Hz, 1H, NH). ^{13}C NMR (75 MHz, CDCl3): δ 18.8, 19.0, 26.4, 29.1, 29.2, 29.3, 30.2, 35.1, 47.9, 78.6, 89.3, 119.5, 124.2, 128.3, 146.2, 154.5, 194.5. HRMS (ESI) calc. for [$C_{18}H_{22}N_3O_5$]$^-$ [M − H]$^-$ 360.1568, found 360.1565.

1-(5-Hydroxy-7,8-dimethyl-5a-nitro-4,5,5a,6,9,9a-hexahydroisoxazolo[4,3-c]isoquinolin-3-yl)-hexan-1-one (**6f**) 35%. Beige powder. M.p. 96–98 °C. ^{1}H NMR (300 MHz, CDCl3): δ 0.92 (t, J = 6.4 Hz, 3H, Me(n-C$_5$H$_{11}$)), 1.33–1.42 (m, 4H, 2CH$_2$, n-C$_5$H$_{11}$), 1.55 (s, 3H, Me), 1.73 (m, 5H, Me+CH$_2$(n-C$_5$H$_{11}$)), 2.42 (d, J = 18.2 Hz, 1H), 2.55–2.99 (m, 5H, 2CH$_2$+CH$_2$(n-C$_5$H$_{11}$)), 3.52 (br.s, 1H, OH), 4.29 (d, J = 6.4 Hz, 1H), 5.34 (d, J = 3.6 Hz, 1H), 6.24 (d, J = 3.3 Hz, 1H, NH). ^{13}C NMR (75 MHz, CDCl3): δ 14.0, 18.7, 19.0, 22.5, 23.5, 29.4, 30.2, 31.6, 35.1, 39.4, 78.7, 89.4, 119.6, 124.2, 128.1, 146.4, 154.6, 192.0. HRMS (ESI) calc. for [$C_{18}H_{26}N_3O_5$]$^+$ [M + H]$^+$ 364.1866, found 364.1875.

(5-Ethoxy-7,8-dimethyl-5a-nitro-4,5,5a,6,9,9a-hexahydroisoxazolo[4,3-c]isoquinolin-3-yl)-(p-tolyl)methanone (**6g**) 42%. Yellowish powder. M.p. 134–136 °C. ^{1}H NMR (300 MHz, CDCl3): δ 1.13 (t, J = 7.0 Hz, 3H, Et), 1.54 (s, 3H, Me), 1.65 (s, 3H, Me), 2.47–3.07 (m, 7H, 2CH$_2$+Me(*p*-Tolyl)), 3.50–3.32 (m, 1H, CH$_2$, Et), 3.84–3.67 (m, 1H, CH$_2$, Et), 4.30 (d, J = 6.8 Hz, 1H), 4.95 (d, J = 4.0 Hz, 1H), 6.78 (d, J = 3.4 Hz, 1H, NH), 7.35 (d, J = 8.1 Hz, 2H, *p*-Tolyl), 8.22 (d, J = 8.2 Hz, 2H, *p*-Tolyl). ^{13}C NMR (75 MHz, CDCl3): δ 14.7, 18.8, 19.0, 22.0, 29.7, 30.2, 35.2, 63.9, 84.5, 88.9, 119.3, 124.2, 129.6, 130.0, 131.4, 133.2, 144.8, 146.9, 154.6, 181.2. HRMS (ESI) calc. for [$C_{22}H_{25}N_3O_5$]$^+$ [M + H]$^+$ 412.1866, found 412.1859.

Supplementary Materials: NMR spectra, HRMS and X-ray analysis data.

Author Contributions: All authors discussed, commented on and wrote the manuscript. All authors have read and agreed to the published version of the manuscript.

Funding: This work was supported by the Russian Science Foundation. Grant 19-73-20259.

Conflicts of Interest: The authors declare no conflict of interest.

References

1. Nepali, K.; Lee, H.-Y.; Liou, J.-P. Nitro-group-containing drugs. *J. Med. Chem.* **2019**, *62*, 2851–2893. [CrossRef] [PubMed]
2. Parry, R.; Nishino, S.; Spain, J. Naturally-occurring nitro compounds. *Nat. Prod. Rep.* **2011**, *28*, 152–167. [CrossRef] [PubMed]
3. Le, S.T.; Asahara, H.; Nishiwaki, N. An alternative synthetic approach to 3-alkylated/arylated 5-nitropyridines. *J. Org. Chem.* **2015**, *80*, 8856–8858. [CrossRef]
4. Yang, S.; Yan, J.; Wang, L.; Guo, X.; Li, S.; Wu, G.; Zuo, R.; Huang, X.; Wang, H.; Wang, L.; et al. Nitropyridinyl Ethyleneimine Compound, the Pharmaceutical Composition Containing It the Preparation Method and Use Thereof. U.S. Patent 2014/0073796 A1, 13 March 2014.
5. Jhons, B.A.; Spaltenstein, A. HIV Integrase Inhibitors. U.S. Patent WO 2007/09101 A2, 18 January 2007.
6. Romero, D.L.; Morge, R.A.; Biles, C.; Discovery, synthesis, and bioactivity of bis(heteroaryl)piperazines. 1. a novel class of non-nucleoside HIV-1 reverse transcriptase inhibitors. *J. Med. Chem.* **1994**, *37*, 999–1014. [CrossRef]
7. Korolev, S.P.; Kondrashina, O.V.; Druzhilovsky, D.S.; Starosotnikov, A.M.; Dutov, M.D.; Bastrakov, M.A.; Dalinger, I.L.; Filimonov, D.A.; Shevelev, S.A.; Poroikov, V.V.; et al. Structural-Functional Analysis of 2,1,3-Benzoxadiazoles and Their N-oxides as HIV-1 Integrase Inhibitors. *Acta Nat.* **2013**, *5*, 63–72. [CrossRef]
8. Kaufman, H.A. Process for Producing Certain Amino Substituted Nitro Pyridines. U.S. Patent 3470172 A, 30 September 1969.
9. Zhang, Q.; Zuo, J.; Chen, X.-L.; Yu, W. Synthesis and application of nitro-pyridine derivatives. *Chem. Res. Appl.* **2002**, *14*, 383–386.
10. Gao, H.; Shreeve, J.M. Azole-based energetic salts. *Chem. Rev.* **2011**, *111*, 7377–7436. [CrossRef]
11. Gryl, M.; Seidler, T.; Wojnarska, J.; Stadnicka, K.; Matulková, I.; Němec, I.; Němec, P. Co-crystals of 2-amino-5-nitropyridine barbital with extreme birefringence and large second harmonic generation effect. *Chem. Eur. J.* **2018**, *24*, 8727–8731. [CrossRef]
12. Terrier, F. *Nucleophilic Aromatic Displacement*; Feuer, H., Ed.; VCH: New York, NY, USA, 1991; pp. 18–138.
13. Chupakhin, O.N.; Charushin, V.N. Recent advances in the field of nucleophilic aromatic substitution of hydrogen. *Tetrahedron Lett.* **2016**, *57*, 2665–2672. [CrossRef]
14. Makosza, M. Reactions of nucleophiles with nitroarenes: Multifacial and versatile electrophiles. *Chem. Eur. J.* **2014**, *20*, 5536–5545. [CrossRef]
15. Makosza, M. Nucleophilic substitution of hydrogen in electron-deficient arenes, a general process of great practical value. *Chem. Soc. Rev.* **2010**, *38*, 2855–2868. [CrossRef] [PubMed]
16. Terrier, F.; Millot, F.; Norris, W.P. Meisenheimer complexes: A kinetic study of water and hydroxide ion attacks on 4,6-dinitrobenzofuroxan in aqueous solution. *J. Am. Chem. Soc.* **1976**, *98*, 5883–5890. [CrossRef]
17. Buncel, E.; Crampton, M.R.; Strauss, M.J.; Terrier, F. *Electron-Deficient Aromatic and Heteroaromatic Base Interactions*; Elsevier: Amsterdam, The Netherlands, 1984; pp. 166–296.
18. Buncel, E.; Dust, J.; Terrier, F. Rationalizing the regioselectivity in polynitroarene anionic. sigma.-adduct formation. Relevance to nucleophilic aromatic substitution. *Chem. Rev.* **1995**, *95*, 2261–2280. [CrossRef]
19. Spear, R.J.; Norris, W.P.; Read, R.W. Direct (uncatalysed) formation of Meisenheimer complexes from primary, secondary and tertiary arylamines. *Tetrahedron Lett.* **1983**, *24*, 1555–1558. [CrossRef]
20. Terrier, F.; Chatrousse, A.P.; Soudais, Y.; Hlaibi, M. Methanol attack on highly electrophilic 4,6-dinitrobenzofurazan and 4,6-dinitrobenzofuroxan derivatives. A kinetic study. *J. Org. Chem.* **1984**, *49*, 4176–4181. [CrossRef]
21. Strauss, M.J.; Renfrow, R.A.; Buncel, E. Ambident aniline reactivity in meisenheimer complex formation. *J. Am. Chem. Soc.* **1983**, *105*, 2473–2474. [CrossRef]

22. Buncel, E.; Renfrow, R.A.; Strauss, M.J. Ambident nucleophilic reactivity in. sigma.-complex formations. 6. Reactivity-selectivity relationships in reactions of ambident nucleophiles with the superelectrophiles 4,6-dinitrobenzofuroxan and 4,6-dinitro-2-(2,4,6-trinitrophenyl) benzotriazole 1-oxide. *J. Org. Chem.* **1987**, *52*, 488–495. [CrossRef]
23. Strauss, M.J.; De Fusco, A.; Terrier, F. Sulfur base complexes of 4,6-dinitrobenzofuroxan. Isolation of cysteine complex; Support for cellular thiol addition as a primary step in the in vitro inhibition of nucleic acid synthesis by nitrobenzofuroxans. *Tetrahedron Lett.* **1981**, *22*, 1945–1948. [CrossRef]
24. Norris, W.P.; Spear, R.J.; Read, R.W. Explosive Meisenheimer complexes formed by addition of nucleophilic reagents to 4,6-dinitrobenzofurazan 1-oxide. *Aust. J. Chem.* **1983**, *36*, 297–309. [CrossRef]
25. Terrier, F.; Halle', J.C.; Pouet, M.J.; Simonnin, M.P. The proton sponge as nucleophile. *J. Org. Chem.* **1986**, *51*, 410–411. [CrossRef]
26. Halle', J.C.; Simonnin, M.P.; Pouet, M.J.; Terrier, F. Nucleophilic displacement of hydrogen in 4,6-dinitro-benzofuroxan and benzofurazan. Synthesis of some novel 7-substituted 4,6-dinitro-benzofuroxans and- benzofurazans. *Tetrahedron Lett.* **1985**, *26*, 1307–1310. [CrossRef]
27. Terrier, F.; Halle', J.C.; Pouet, M.J.; Simonnin, M.P. Nonconventional electrophilic heteroaromatic substitutions: Ring vs. side-chain reactivity of 2,5-dimethyl five-membered ring heterocycles toward electron-deficient aromatics. *J. Org. Chem.* **1984**, *49*, 4363–4367. [CrossRef]
28. Terrier, F.; Kizilian, E.; Halle', J.C.; Buncel, E. 4,6-dinitrobenzofuroxan: A stronger electrophile than the p-nitrobenzenediazonium cation and proton. *J. Am. Chem. Soc.* **1992**, *114*, 1740–1742. [CrossRef]
29. Terrier, F.; Pouet, M.J.; Halle', J.C.; Hunt, S.; Jones, J.R.; Buncel, E. Electrophilic heteroaromatic substitutions: Reactions of 5-X-substituted indoles with 4,6-dinitrobenzofuroxan. *J. Chem. Soc. Perkin Trans.* **1993**, *2*, 16651–16672. [CrossRef]
30. Terrier, F.; Pouet, M.J.; Halle', J.C.; Kizilian, E.; Buncel, E. Electrophilic aromatic substitutions: Reactions of hydroxyl- and methoxy-substituted benzenes with 4.6-dinitrobenzofuroxan: Kinetics and mechanism. *J. Phys. Org. Chem.* **1998**, *11*, 707–714. [CrossRef]
31. Kind, J.; Niclas, H. Preparation of 5,7-dinitrobenzofurazan-4-yl aryl diketones via meisenheimer complexes. *J. Synth. Commun.* **1993**, *23*, 1569–1576. [CrossRef]
32. Lakhdar, S.; Goumont, R.; Boubaker, T.; Mokhtari, M.; Terrier, F. Nitrobenzoxadiazoles and related heterocycles: A relationship between aromaticity, superelectrophilicity and pericyclic reactivity. *Org. Biomol. Chem.* **2006**, *4*, 1910–1919. [CrossRef]
33. Terrier, F.; Simonnin, M.P.; Pouet, M.J.; Strauss, M.J. Reactivity of carbon acids toward 4,6-dinitrobenzofuroxan. Studies of keto-enol equilibriums and diastereoisomerism in carbon-bonded anionic sigma. Complexes. *J. Org. Chem.* **1981**, *46*, 3537–3543. [CrossRef]
34. Terrier, F.; Sebban, M.; Goumont, R.; Halle', J.C.; Moutiers, G.; Cangelosi, I.; Buncel, E. Dual behavior of 4-aza-6-nitrobenzofuroxan. A powerful electrophile in hydration and σ-complex formation and a potential dienophile or heterodiene in diels-alder type reactions. *J. Org. Chem.* **2000**, *65*, 7391–7398. [CrossRef]
35. Remennikov, G.Y.; Pirozhenko, V.V.; Vdovenko, S.I.; Kravchenko, S.A. σ-Complexes in the pyrimidine series. 13. Reaction of 7- and 5-methoxyfurozano-[3,4-*d*]pyrimidines with some C-nucleophiles. *Chem. Heterocycl. Compd.* **1998**, *34*, 104–110. [CrossRef]
36. Wertjes, W.C.; Southgate, E.H.; Sarlah, D. Recent advances in chemical dearomatization of nonactivated arenes. *Chem. Soc. Rev.* **2018**, *47*, 7996–8017. [CrossRef] [PubMed]
37. Ortiz, F.L.; Iglesias, M.J.; Fernandez, I.; Andujar Sanchez, C.M.; Gomez, G.R. Nucleophilic dearomatizing (DNAr) reactions of aromatic, C,H-Systems. A mature paradigm in organic synthesis. *Chem. Rev.* **2007**, *107*, 1580–1691. [CrossRef] [PubMed]
38. Bastrakov, M.A.; Nikol'skiy, V.V.; Starosotnikov, A.M.; Fedyanin, I.V.; Shevelev, S.A.; Knyazev, D.A. Reactions of 3-R-5-nitropyridines with nucleophiles: Nucleophilic substitution vs. conjugate addition. *Tetrahedron* **2019**, *75*, 130659. [CrossRef]
39. Konstantinova, L.S.; Bastrakov, M.A.; Starosotnikov, A.M.; Glukhov, I.V.; Lysov, K.A.; Rakitin, O.A.; Shevelev, S.A. 4.6-dinitrobenzo[*c*]isothiazole: Synthesis and 1,3-dipolar cycloaddition to azomethine ylide. *Mendeleev Commun.* **2010**, *20*, 353–354. [CrossRef]
40. Bastrakov, M.A.; Leonov, A.I.; Starosotnikov, A.M.; Fedyanin, I.V.; Shevelev, S.A. 8-R-5,7-dinitroquinolines in [3 + 2]-cycloaddition reactions with N-methylazomethine ylide. *Russ. Chem. Bull. Int. Ed.* **2013**, *62*, 1052–1059. [CrossRef]

41. Bastrakov, M.A.; Starosotnikov, A.M.; Fedyanin, I.V.; Kachala, V.V.; Shevelev, S.A. 5-nitro-7,8-furoxanoquinoline: A new type of fused nitroarenes possessing Diels-Alder reactivity. *Mendeleev Commun.* **2014**, *24*, 203–205. [CrossRef]
42. Bastrakov, M.A.; Starosotnikov, A.M.; Pavlov, A.A.; Dalinger, I.L.; Shevelev, S.A. Synthesis of novel polycyclic heterosystems from 5-nitro[1,2,5]selenadiazolo[3,4-*e*]benzofuroxans. *Chem. Heterocycl. Compd.* **2016**, *52*, 690–693. [CrossRef]
43. Starosotnikov, A.M.; Bastrakov, M.A.; Pavlov, A.A.; Fedyanin, I.V.; Dalinger, I.L.; Shevelev, S.A. Synthesis of novel polycyclic heterosystems based on 5-nitro[1,2,5]thiadiazolo[3,4-*e*]benzofuroxan. *Mendeleev Commun.* **2016**, *26*, 217–219. [CrossRef]
44. Bastrakov, M.A.; Kucherova, A.Y.; Fedorenko, A.K.; Starosotnikov, A.M.; Fedyanin, I.V.; Dalinger, I.L.; Shevelev, S.A. Dearomatization of 3,5-dinitropyridines—An atom-efficient approach to fused 3-nitropyrrolidines. *Arkivoc* **2017**. [CrossRef]
45. Starosotnikov, A.M.; Shkaev, D.V.; Bastrakov, M.A.; Fedyanin, I.V.; Shevelev, S.A.; Dalinger, I.L. Nucleophilic dearomatization of 4-aza-6-nitrobenzofuroxan by CH acids in the synthesis of pharmacology-oriented compounds. *Beilstein J. Org. Chem.* **2017**, *13*, 2854–2861. [CrossRef]
46. Starosotnikov, A.M.; Shkaev, D.V.; Bastrakov, M.A.; Shevelev, S.A.; Dalinger, I.L. Dearomatization of oxa- or selenadiazolopyridines with neutral nucleophiles as an efficient approach to pharmacologically relevant nitrogen compounds. *Mendeleev Commun.* **2018**, *28*, 638–640. [CrossRef]
47. Bastrakov, M.A.; Fedorenko, A.K.; Starosotnikov, A.M.; Kachala, V.V.; Shevelev, S.A. Dearomative (3 + 2) cycloaddition of 2-substituted 3,5-dinitropyridines and *N*-methyl azomethine ylide. *Chem. Heterocycl. Compd.* **2019**, *55*, 72–77. [CrossRef]
48. Starosotnikov, A.M.; Il'kov, K.V.; Bastrakov, M.A.; Fedyanin, I.V.; Kokorekin, V.A. Mild and efficient addition of carbon nucleophiles to condensed pyridines: Influence of structure and limits of applicability. *Chem. Heterocycl. Comp.* **2020**, *56*, 92–100. [CrossRef]
49. Cikotiene, I. Intramolecular iodine-mediated oxygen transfer from nitro groups to C ≡ C Bonds. *Eur. J. Org. Chem.* **2012**, 27662–27773. [CrossRef]
50. Bruno, I.J.; Cole, J.C.; Kessler, M.; Luo, J.; Motherwell, W.D.S.; Purkis, L.H.; Smith, B.R.; Taylor, R.; Cooper, R.I.; Harris, S.E.; et al. Retrieval of crystallographically-derived molecular geometry information. *J. Chem. Inf. Comput. Sci.* **2004**, *44*, 2133–2144. [CrossRef]
51. *SAINT*; Version 8.34A; Bruker AXS Inc.: Madison, WI, USA, 2014.
52. Krause, L.; Herbst-Irmer, R.; Sheldrick, G.M.; Stalke, D. Comparison of silver and molybdenum microfocus X-ray sources for single-crystal structure determination. *J. Appl. Cryst.* **2015**, *48*, 3–10. [CrossRef]
53. Sheldrick, G.M. SHELXT—Integrated space-group and crystalstructure determination. *Acta Crystallogr. A* **2015**, *71*, 3–8. [CrossRef]
54. Sheldrick, G.M. Crystal structure refinement with SHELXL. *Acta Crystallogr. C* **2015**, *71*, 3–8. [CrossRef]

Sample Availability: Samples of the compounds are not available from the authors.

© 2020 by the authors. Licensee MDPI, Basel, Switzerland. This article is an open access article distributed under the terms and conditions of the Creative Commons Attribution (CC BY) license (http://creativecommons.org/licenses/by/4.0/).

Article

Comparison of Substituting Ability of Nitronate versus Enolate for Direct Substitution of a Nitro Group

Yusuke Mukaijo [1], Soichi Yokoyama [1,2,3] and Nagatoshi Nishiwaki [1,2,*]

1. School of Environmental Science and Engineering, Kochi University of Technology, Tosayamada, Kami, Kochi 782-8502, Japan; 235030n@gs.kochi-tech.ac.jp (Y.M.); yokoyama.soichi@sanken.osaka-u.ac.jp (S.Y.)
2. Research Center for Molecular Design, Kochi University of Technology, Tosayamada, Kami, Kochi 782-8502, Japan
3. The Institute of Scientific and Industrial Research, Osaka University, Mihogaoka, Ibaraki, Osaka 567-0047, Japan
* Correspondence: nishiwaki.nagatoshi@kochi-tech.ac.jp; Tel.: +81-887-57-2517

Academic Editors: Mário J. F. Calvete and Fawaz Aldabbagh
Received: 9 April 2020; Accepted: 23 April 2020; Published: 28 April 2020

Abstract: α-Nitrocinnamate underwent the conjugate addition of an active methylene compound such as nitroacetate, 1,3-dicarbonyl compound, or α-nitroketone, and the following ring closure afforded functionalized heterocyclic frameworks. The reaction of cinnamate with nitroacetate occurs via nucleophilic substitution of a nitro group by the *O*-attack of the nitronate, which results in isoxazoline *N*-oxide. This protocol was applicable to 1,3-dicarbonyl compounds to afford dihydrofuran derivatives, including those derived from direct substitution of a nitro group caused by *O*-attack of enolate. It was found the reactivity was lowered by an electron-withdrawing group on the carbonyl moiety. When α-nitroketone was employed as a substrate, three kinds of products were possibly formed; of these, only isoxazoline *N*-oxide was identified. This result indicates that the substituting ability of nitronate is higher than that of enolate for the direct S_N2 substitution of a nitro group.

Keywords: conjugate addition; dihydrofuran; 1,3-dicarbonyl compound; enolate; isoxazoline *N*-oxide; nitro group; nitroketone; nitronate; nucleophilic substitution

1. Introduction

The nitro group is one of the important functional groups because of its unique chemical properties, which are useful in many compounds. The strong electron-withdrawing property of the nitro group reduces the electron density of the adjacent atom or double bond through both inductive and resonance effects. The increased electrophilicity facilitates nucleophilic addition to a nitroalkene, while the resulting anionic intermediate is stabilized by the nitro group. The nitro group also serves as a good leaving group. Nitroalkane undergoes E2 elimination of nitrous acid to afford C–C double bonds under basic [1–4] or acidic conditions [5,6]. Direct S_N2 substitution is also sometimes observed [7], in which a nitrite ion is eliminated. The high acidity of an α-proton of the nitro group easily generates a nitronate, which possesses both nucleophilic [8,9] and electrophilic sites to serve as a 1,3-dipole [10,11]. Moreover, the nitro group is a precursor of amino and carbonyl groups by reduction and Nef-type reactions [12], respectively. The many properties of the nitro group have facilitated diverse applications. Recently, the complex/mixed properties of the nitro group have attracted considerable attention for the synthesis of multi-functionalized/polyfunctionalized compounds. In our previous work, α-nitrocinnamate served as a precursor of functionalized enynes via conjugate addition of an acetylide ion followed by elimination of a nitrous acid [1]. When α-chloro-α,β-unsaturated ketone is subjected to the reaction with cyano-*aci*-nitroacetate, intramolecular nucleophilic substitution of the chloro group by the nitronate ion occurs after Michael addition [13]. Based on these works, this study furthered this topic by studying the synthesis of functionalized heterocyclic compounds using a combination of conjugate addition of an active methylene compound and subsequent O-attack of the resulting nitronate/enolate which undergoes direct substitution of the nitro group. The substituting abilities of the nitronate/enolate were compared.

2. Results and Discussion

As a highly electron-deficient substrate, methyl α-nitrocinnamate (**1a**) was employed because of its easy availability via condensation of benzaldehyde and nitroacetate in the presence of piperidine hydrochloride with the removal of water as an azeotrope mixture. Indeed, the reaction of **1a** with propylamine quantitatively proceeded at room temperature in acetonitrile to afford product **2a** as a 1:1 mixture of diastereomers, which implies that two stereocenters were newly formed; however, the product was not an adduct of **1a** and propylamine. Based on spectral data of the reaction mixture, compound **2a** was confirmed to be an adduct of **1a** with methyl nitroacetate generated in situ. This reaction is thought to proceed as shown in Scheme 1. After conjugate addition of propylamine to **1a**, C–C bond cleavage forms nitroacetate [14]. Indeed, signals of 1-phenyl-*N*-propylmethanimine (**3**) were observed in the ^{1}H NMR spectrum of the reaction mixture (see Supplementary Materials). The generated nitroacetate underwent the conjugate addition to another cinnamate **1a** to afford product **2a**. However, product **2a** could not be isolated by column chromatography because of its instability on silica gel. On the other hand, when the mixture of **2a** and **3** was left at room temperature without solvent for 7 days, ring closure proceeded to afford isoxazoline *N*-oxide **4a**, in which the nitronate underwent the direct substitution of the nitro group to form an isoxazoline framework (Scheme 1).

There have been several studies of the formation of isoxazoline *N*-oxides from α-nitro-α,β-unsaturated esters with C–H acids such as secondary nitroalkane [15], (ethoxycarbonylmethyl)dimethylsulfonium salt [16], (ethoxycarbonylmethyl)ammonium salt [17], (ethoxycarbonylmethyl)pyridinium salt [18], α-halomalonate [19], and α-iodo aldehyde [20]. β,β-Dimethoxynitroethene is also usable as a nucleophile in this protocol [21]. Among these, only two methods employ a nitro group as a leaving group [15,21]. In these reactions, the nucleophilicity of the nitronate ion is relatively high. To the contrary, nucleophilicity of the nitronate in **2a** is considered to be lower due to the electron-withdrawing ester functionality.

Thus, ethyl α-nitrocinnamate (**1b**) was allowed to react with ethyl nitroacetate in the presence of triethylamine. It was confirmed that the successive conjugate-addition/ring-closure reactions efficiently proceeded in one pot to afford isoxazoline *N*-oxide **4b** [22] (Scheme 2). This result prompted the study of α-nitrocinnamate **1b** reactions with other active methylene compounds such as 1,3-dicarbonyl compounds, because the nucleophilicity of the enolate ion is considered to be lower than that of the nitronate ion.

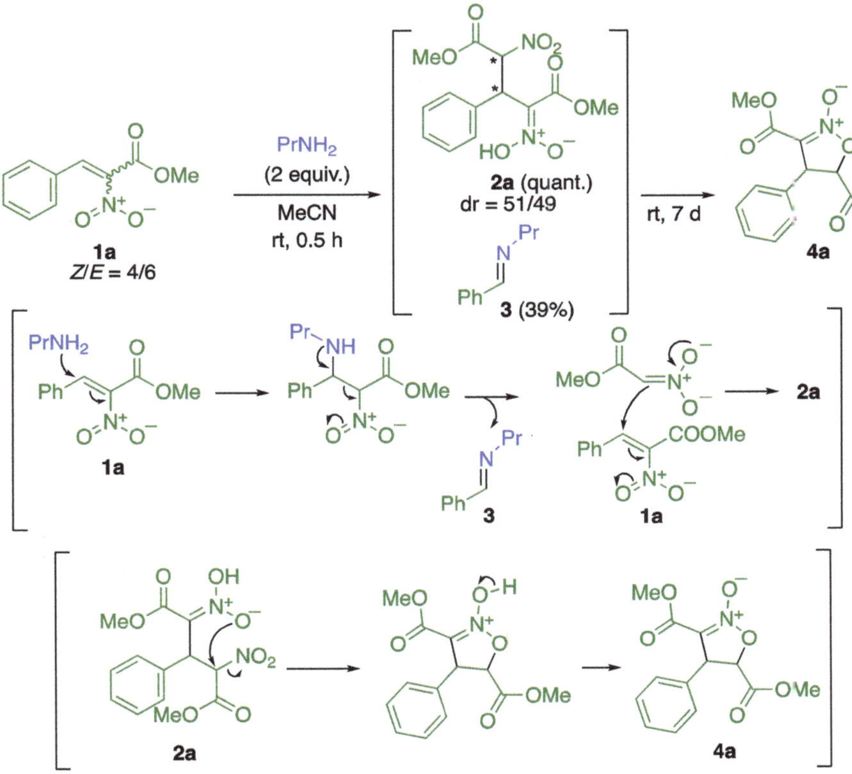

Scheme 1. Reaction of α-nitrocinnamate **1a** with propylamine and the subsequent ring closure, with plausible mechanisms.

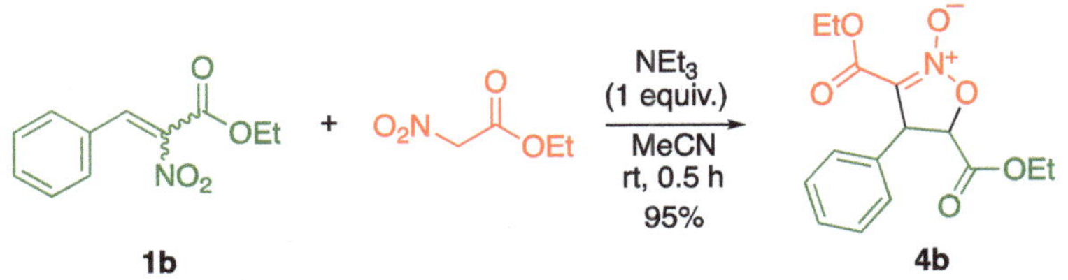

Scheme 2. Synthesis of isoxazoline *N*-oxide **4b**.

Although two studies on the reactions of **1b** with β-keto esters were found in the literature [23,24], they were not conducted under same conditions (one reaction was conducted in the presence of tetrabutylammonium bromide). In order to compare the reactivity systematically and to generalize this protocol, the same reaction conditions should be used. Therefore, 1,3-dicarbonyl compounds **5a–f** were subjected to reactions with **1b** in the presence of triethylamine at 60 °C for 3 h in acetonitrile (Table 1). Cyclization efficiently proceeded to produce a furan derivative **6a** [23] with 92% yield when ethyl acetoacetate was employed as a substrate (entry 1). The nucleophilicity of the enolate of the ketone functionality was higher than that of the ester functionality because of the electron-withdrawing inductive effect of the ethoxy group. Next, the acetyl group of **5a** was replaced with a trifluoroacetyl group. In this case, decreasing the nucleophilicity of the enolate ion was more effective than increasing the acidity of the methylene group, which produced 5-(trifluoromethyl)-2,3-dihydrofuran **6b** with a lower yield (entry 2). When benzoylacetate **5c** was reacted under the same conditions, the cyclization occurred without significant effects from the steric hindrance of the phenyl group to furnish the corresponding dihydrofuran **6c** (entry 3). Diketone **5d**, acetylacetone, exhibited higher reactivity than keto esters **5a–c**, and yielded **6d** [24] quantitatively (entry 4). Cyclic diketone **5e**, 1,3-cyclohexanedione, also underwent the reaction efficiently to produce bicyclic furan **6e**, which was not influence by steric strain (entry 5) [23,25]. On the other hand, the formation of ester-substituted furan **6f** was not detected when diester **5f** was subjected to the reaction conditions because of the low nucleophilicity of the enolate ion (entry 6).

Table 1. Synthesis of functionalized 2,3-dihydrofuran **6**.

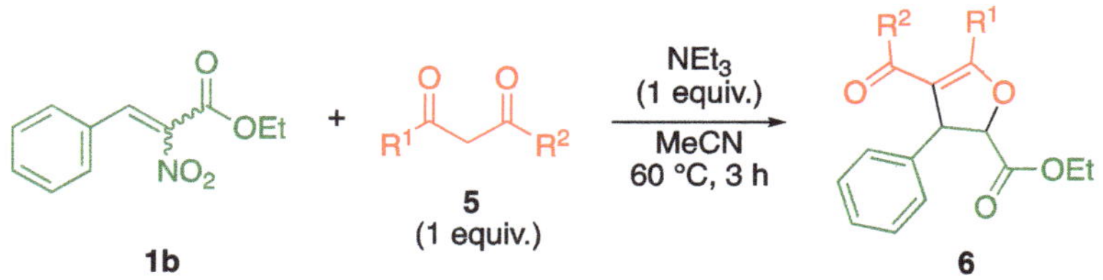

Entry	R^1	R^2	5 and 6	Yield/%
1	Me	OEt	a	92
2	CF_3	OEt	b	43
3	Ph	OEt	c	62
4	Me	Me	d	quant.
5	$-(CH_2)_3-$		e	quant.
6	OEt	OEt	f	0

In the reaction of cinnamate **1b** and diketone **5d**, the conjugate addition of enolate ion of **5d** to **1b** afforded adduct intermediate **8d**, from which 2,3-dihydrofuran **6d** was formed by the substitution of the nitro group. In other words, 2,3-dihydrofuran **6d** was also synthesized if intermediate **8d** was formed [24]. Indeed, when diacetylated styrene **7** was reacted with ethyl nitroacetate in the presence of triethylamine under the same conditions, a high yield of dihydrofuran **6d** was obtained (Scheme 3).

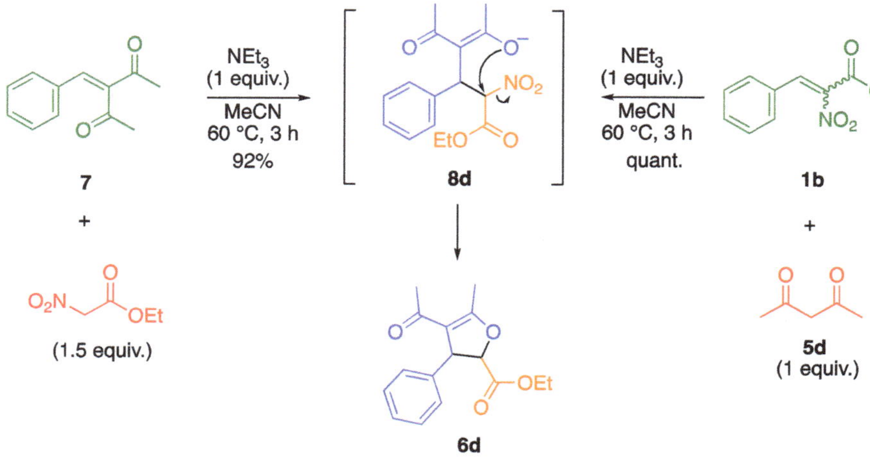

Scheme 3. Synthesis of 2,3-dihydrofuran **6d** from either diacetylated styrene **7** or cinnamate **1b**.

Next, α-nitroketone **9** was employed as a nucleophile able to produce both an enolate and a nitronate ion, which facilitated the comparison of the substituting ability of these anions directly (Scheme 4). In this case, adduct **10** is thought to have formed intermediately, from which three possible structures **11**–**13** could be produced. Dihydrofuran **11** was formed by the attack of enolate (Path a), and isoxazoline *N*-oxide **12** is formed by the attack of nitronate derived from nitroketone (Path b). On the other hand, attack of the nitronate derived from nitrocinnamate affords isoxazoline *N*-oxide **13** (Path c).

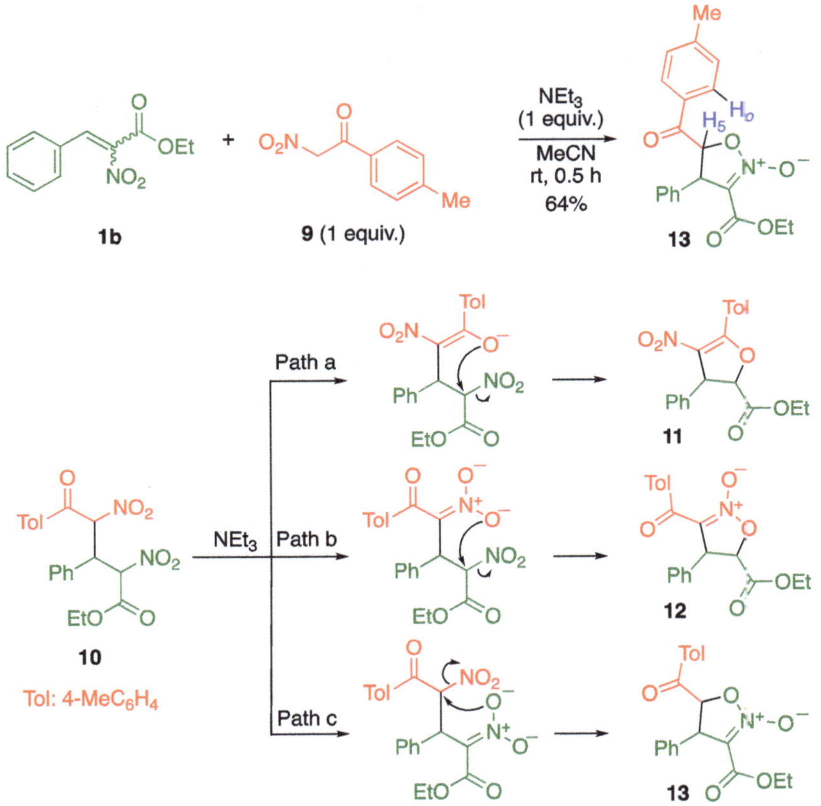

Scheme 4. Reaction of cinnamate **1b** with nitroketone **9** and three possible products **11**–**13**.

When cinnamate **1b** was allowed to react with nitroketone **9** in the presence of triethylamine, the reaction mixture became somewhat complex, so only one cyclic product was isolated as a major product. Since a lot of small signals were observed in the ^{1}H NMR spectrum of the reaction mixture, it is difficult to know whether cyclic products were formed as minor products or not. In the ^{13}C NMR spectrum of the major product, a signal corresponding to carbonyl carbon was observed at 191 ppm, which indicated that the product had a ketone functionality; thus, the possibility of **11** was excluded. In the ^{1}H-^{1}H NOESY 2D NMR spectrum, a correlation was observed between the proton at the 5-position of the isoxazoline ring and the *ortho*-proton of the benzoyl group, by which the product was determined to be isoxazoline *N*-oxide **13**.

This result indicated that the nitronate ion substituted a nitro group via Path c. It is considered that the different reactivity of the two nitro groups was caused by the different electron-withdrawing ability of the ketone and the ester functionalities. The stronger electron-withdrawing toluoyl group increased the electrophilicity of the α-carbon and decreased the nucleophilicity of the nitronate ion, which facilitated the reaction via Path c leading to the predominant formation of **13**.

3. Experimental Section

3.1. General

All reagents were purchased from commercial sources and used without further purification. ^{1}H and ^{13}C NMR spectra were recorded on Bruker DPX-400 spectrometer (400 MHz and 100 MHz, respectively, Billerica, MA, USA) in CDCl$_3$ using TMS as an internal standard. The ^{13}C NMR assignments were performed via DEPT experiments. A Shimadzu IR spectrometer equipped with an ATR detector (Kyoto, Japan) was used to record infrared spectra. High-resolution mass spectra were obtained on an AB SCIEX Triplet TOF 4600 mass spectrometer (Framingham, MA, USA). Melting points were recorded on a Stanford Research Systems Optimelt automated melting point system (Sunnyvale, CA, USA) and were uncorrected.

3.2. Synthesis of Isoxazoline N-oxide **4b**

To a solution of ethyl α-nitrocinnamate **1b** (94.6 mg, 0.43 mmol) in acetonitrile (1.3 mL), ethyl nitroacetate (48 µL, 0.43 mmol) and triethylamine (60 µL, 0.43 mmol) were added, and the resultant mixture was stirred at room temperature for 30 min. After removal of the solvent under reduced pressure, the residual brown oil was dissolved in ethyl acetate (10 mL) and washed with water (10 mL × 4), and then dried over magnesium sulfate. After removal of the solvent, the residue was purified by column chromatography on silica gel to afford isoxazoline N-oxide **4b** (eluted with hexane/EtOAc = 8/2, 121 mg, 0.41 mmol, 95%) as a yellow solid.

3,5-Bis(ethoxycarbonyl)-4-phenyl-2-isoxazoline 2-oxide (**4b**) [22]. Yellow solid, yield; 95%, m.p. 75–76 °C. ^{1}H NMR (400 MHz, CDCl$_3$) δ 7.40–7.30 (m, 5H), 4.93 (d, J = 2.8 Hz, 1H), 4.84 (d, J = 2.8 Hz, 1H), 4.35 (dq, J = 10.8, 7.2 Hz, 1H), 4.32 (dq, J = 10.8, 7.2 Hz, 1H), 4.21 (dq, J = 10.8, 7.2 Hz, 1H), 4.17 (dq, J = 10.8, 7.2 Hz, 1H), 1.35 (dd, J = 7.2, 7.2 Hz, 3H), 1.17 (dd, J = 7.2, 7.2 Hz, 3H); ^{13}C NMR (100 MHz, CDCl3) δ 168.2 (C), 158.1 (C), 138.1 (C), 129.3 (CH), 128.7 (CH), 127.0 (CH), 109.0 (C), 78.8 (CH), 62.7 (CH2), 62.0 (CH2), 52.7 (CH), 14.1 (CH3), 13.9 (CH3).

3.3. Typical Procedure for Synthesis of 2,3-Dihydrofuran **6**

To a solution of ethyl α-nitrocinnamate **1b** (71.5 mg, 0.32 mmol) in acetonitrile (1.0 mL), ethyl acetoacetate **5a** (41 µL, 0.32 mmol) and triethylamine (45 µL, 0.32 mmol) were added, and the resultant mixture was heated at 60 °C for 3 h. After removal of the solvent under reduced pressure, the residual orange oil was purified by column chromatography on silica gel to afford 2,3-dihydrofuran **6a** (eluted with hexane/EtOAc = 1/1, 88 mg, 0.29 mmol, 92%) as a pale yellow oil. When other 1,3-dicarbonyl compounds **5** were used, the experiment was conducted in the same way.

2,4-Bis(ethoxycarbonyl)-2,3-dihydro-5-methyl-3-phenylfuran (**6a**) [23]. Pale yellow oil, yield; 92%. ^{1}H NMR (400 MHz, CDCl$_3$) δ 7.34–7.30 (m, 2H), 7.27–7.22 (m, 3H), 4.83 (d, J = 4.8 Hz, 1H), 4.41 (dq, J = 4.8, 1.2 Hz, 1H), 4.30 (dq, J = 10.8, 7.2 Hz, 1H), 4.26 (dq, J = 10.8, 7.2 Hz, 1H), 4.04 (dq, J = 10.8, 7.2 Hz, 1H), 3.98 (dq, J = 10.8, 7.2 Hz, 1H), 2.40 (d, J = 1.2 Hz, 3H), 1.32 (dd, J = 7.2, 7.2 Hz, 3H), 1.07 (dd, J = 7.2, 7.2 Hz, 3H); ^{13}C NMR (100 MHz, CDCl$_3$) δ 170.1 (C), 168.4 (C), 164.9 (C), 142.6 (C), 128.6 (CH), 127.2 (CH), 127.1(CH), 106.4 (C), 85.8 (CH), 61.8 (CH2), 59.6 (CH2), 52.8 (CH), 14.2 (CH3), 14.1 (CH3), 14.1 (CH3); IR (ATR/cm^{-1}) 1755, 1701, 1651, 1207, 1088, 1038; HRMS (ESI/TOF) calcd. for [M + H]$^+$ $C_{17}H_{21}O_5$: 305.1384, found: 305.1384.

2,4-Bis(ethoxycarbonyl)-5-trifluoromethyl-2,3-dihydro-3-phenylfuran (**6b**). Yellow oil, yield; 43%. ^{1}H NMR (400 MHz, CDCl$_3$) δ 7.4–7.2 (m, 5H), 4.99 (d, J = 4.8 Hz, 1H), 4.62 (dq, J = 4.8, 2.4 Hz, 1H), 4.32 (dq, J = 11.6, 7.2 Hz, 1H), 4.30 (dq, J = 11.6, 7.2 Hz, 1H), 4.12 (dq, J = 10.8, 7.2 Hz, 1H), 4.05 (dq, J = 10.8, 7.2 Hz, 1H), 1.33 (dd, J = 7.2, 7.2 Hz, 3H), 1.13 (dd, J = 7.2, 7.2 Hz, 3H); ^{13}C NMR (100 MHz, CDCl$_3$) δ

168.5 (C), 161.2 (C), 151.2 (C, q, J = 40.0 Hz), 139.8 (C), 129.1 (CH), 128.1 (CH), 127.2 (CH), 118.0 (C, q, J = 271.0 Hz), 113.0 (C, q, J = 3.0 Hz), 86.2 (CH), 62.3 (CH_2), 61.1 (CH_2), 53.8 (CH), 14.1 (CH_3), 13.7 (CH_3); IR (ATR/cm^{-1}) 1759, 1728, 1200, 1157, 1111; HRMS (ESI/TOF) calcd. for $[M + H]^+$ $C_{17}H_{18}F_3O_5$: 359.1101, found: 359.1092.

2,4-Bis(ethoxycarbonyl)-2,3-dihydro-1,3-diphenylfuran (**6c**). Colorless oil, yield; 62%. ^{1}H NMR (400 MHz, CDCl$_3$) δ 7.98–7.95 (m, 2H), 7.48–7.42 (m, 3H), 7.42–7.35 (m, 4H), 7.35–7.28 (m, 1H), 4.96 (d, J = 4.4 Hz, 1H), 4.62 (d, J = 4.4 Hz, 1H), 4.34 (dq, J = 10.8, 7.2 Hz, 1H), 4.31 (dq, J = 10.8, 7.2 Hz, 1H), 4.01 (dq, J = 10.8, 7.2 Hz, 1H), 3.98 (dq, J = 10.8, 7.2 Hz, 1H), 1.36 (dd, J = 7.2, 7.2 Hz, 3H), 1.03 (dd, J = 7.2, 7.2 Hz, 3H); ^{13}C NMR (100 MHz, CDCl$_3$) δ 170.2 (C), 165.4 (C), 164.1 (C), 142.5 (C), 130.1 (CH), 129.8 (CH), 129.2 (C), 128.8 (CH), 127.7 (CH), 127.4 (CH), 127.2 (CH), 106.7 (C), 85.0 (CH), 61.8 (CH_2), 59.9 (CH_2), 54.1 (CH), 14.2 (CH_3), 13.9 (CH_3); IR (ATR/cm^{-1}) 1751, 1697, 1628, 1203, 1076, 752, 694; HRMS (ESI/TOF) calcd. for $[M + H]^+$ $C_{22}H_{23}O_5$: 367.1540, found: 367.1540.

4-Ethanoyl-2-ethoxycarbonyl-2,3-dihydro-5-methyl-3-phenylfuran (**6d**) [24]. Yellow solid, yield; quant., m.p. 63–64 °C. ^{1}H NMR (400 MHz, CDCl$_3$) δ 7.37–7.33 (m, 2H), 7.30–7.23 (m, 3H), 4.72 (d, J = 4.8 Hz, 1H), 4.49 (dq, J = 4.8, 1.2 Hz, 1H), 4.31 (dq, J = 10.8, 7.2 Hz, 1H), 4.27 (dq, J = 10.8, 7.2 Hz, 1H), 2.44 (d, J = 1.2 Hz, 3H), 1.95 (s, 3H), 1.34 (dd, J = 7.2, 7.2 Hz, 3H); ^{13}C NMR (100 MHz, CDCl$_3$) δ 194.3 (C), 170.0 (C), 168.6 (C), 142.2 (C), 129.1 (CH), 127.6 (CH), 127.2 (CH), 115.1 (C), 86.0 (CH), 61.9 (CH_2), 53.3 (CH), 29.6 (CH_3), 14.9 (CH_3), 14.2 (CH_3); IR (ATR/cm^{-1}) 1755, 1674, 1624, 1604, 1196, 1038; HRMS (ESI/TOF) calcd. for $[M + H]^+$ $C_{16}H_{18}O_4Na$: 297.1097, found: 297.1099.

5,6-Cyclohexa-2-ethoxycarbonyl-2,3-dihydro-3-phenylfuran-4-one (**6e**) [25]. Yellow oil, yield; quant. ^{1}H NMR (400 MHz, CDCl$_3$) δ 7.34–7.31 (m, 2H), 7.27–7.21 (m, 3H), 4.96 (d, J = 4.8 Hz, 1H), 4.46 (br d, J = 4.8 Hz, 1H), 4.32 (dq, J = 10.8, 7.2 Hz, 1H), 4.27 (dq, J = 10.8, 7.2 Hz, 1H), 2.68–2.65 (m, 2H), 2.44–2.31 (m, 2H), 2.19–2.10 (m, 2H), 1.33 (dd, J = 7.2, 7.2 Hz, 3H); ^{13}C NMR (100 MHz, CDCl$_3$) δ 194.3 (C), 177.4 (C), 169.5 (C), 141.1 (C), 128.9 (CH), 127.4 (CH), 127.0 (CH), 115.8 (C), 88.0 (CH), 62.0 (CH_2), 49.8 (CH), 36.8 (CH_2), 23.9 (CH_2), 21.7 (CH_2), 14.2 (CH_3); IR (ATR/cm^{-1}) 1751, 1639, 1396, 1219, 748; HRMS (ESI/TOF) calcd. for $[M + H]^+$ $C_{17}H_{19}O_4$: 287.1278, found: 287.1278.

3-Ethoxycarbonyl-4,5-dihydro-5-(4-methylbenzoyl)-4-phenylisoxazoline 2-oxide (**13**). Yellow oil, yield; 64%. ^{1}H NMR (400 MHz, CDCl$_3$) δ 7.81 (d, J = 8.0 Hz, 2H), 7.42–7.36 (m, 5H), 7.29 (d, J = 8.0 Hz, 2H), 5.66 (d, J = 3.6 Hz, 1H), 5.12 (d, J = 3.6 Hz, 1H), 4.19 (dq, J = 10.8, 7.2 Hz, 1H), 4.14 (dq, J = 10.8, 7.2 Hz, 1H), 2.43 (s, 3H), 1.13 (dd, J = 7.2, 7.2 Hz, 3H); ^{13}C NMR (100 MHz, CDCl$_3$) δ 191.3 (C), 158.4 (C), 146.0 (C), 138.6 (C), 130.9 (C), 129.9 (CH), 129.5 (CH), 129.5 (CH), 128.8 (CH), 127.6 (CH), 109.8 (C), 81.7 (CH), 62.0 (CH_2), 51.8 (CH), 22.0 (CH_3), 14.0 (CH_3); IR (KBr/cm^{-1}) 1736, 1697, 1628, 1606, 1228, 740; HRMS (ESI/TOF) calcd. for $[M + H]^+$ $C_{20}H_{20}NO_5$: 354.1336 found: 354.1337.

4. Conclusions

2,3-Dihydrofurans and isoxazoline *N*-oxides were synthesized from α-nitrocinnamate **1** and active methylene compounds by conjugate addition and the subsequent *O*-attack. Via a series of reactions using several substrates, the nitro group increased the electrophilicity of the α-carbon and served as a good leaving group. The nitro group also served as a good nucleophile when it was converted to nitronate ion, which is more reactive than the enolate ion of ketone or ester functionalities. These results will be useful information for researchers studying synthetic chemistry using the multi-functionalities of a nitro group.

Supplementary Materials: The following are available online. ^{1}H and ^{13}C NMR spectra of **2a, 4, 6** and **13**.

Author Contributions: Y.M. did experiments and wrote a draft; S.Y. and N.N. analyzed data and discussed with Y.M.; all authors contributed to the revision. All authors have read and agreed to the published version of the manuscript.

Funding: This research received no external funding.

Conflicts of Interest: The authors declare no conflict of interest.

References

1. Asahara, H.; Sofue, A.; Kuroda, Y.; Nishiwaki, N. Alkynylation and cyanation of alkenes using diverse properties of a nitro group. *J. Org. Chem.* **2018**, *83*, 13691–13699. [CrossRef]
2. Nishiwaki, N. Chemistry of nitroquinolones and synthetic application to unnatural 1-methyl-2-quinolone derivatives. *Molecules* **2010**, *15*, 5174–5195. [CrossRef]
3. Hao, F.; Nishiwaki, N. Recent progress in nitro-promoted direct functionalization of pyridones and quinolones. *Molecules* **2020**, *25*, 673. [CrossRef]
4. Chiurchiù, E.; Gabrielli, S.; Ballini, R.; Palmieri, A. A new valuable synthesis of polyfunctionalized furans starting from β-nitroenones and active methylene compounds. *Molecules* **2019**, *24*, 4575. [CrossRef] [PubMed]
5. Palmieri, A.; Gabrielli, S.; Ballini, R. Efficient two-step sequence for the synthesis of 2,5-disubstituted furan derivatives from functionalized nitroalkanes: successive amberlyst A21- and amberlyst 15-catalyzed processes. *Chem. Commun.* **2010**, *46*, 6165–6167. [CrossRef] [PubMed]
6. Palmieri, A.; Gabrielli, S.; Parlapiano, M.; Ballini, R. One-pot synthesis of alkyl pyrrole-2-carboxylates starting from β-nitroacrylates and primary amines. *RSC Adv.* **2015**, *5*, 4210–4213. [CrossRef]
7. Mo, Y.; Liu, S.; Liu, Y.; Ye, L.; Shi, Z.; Zhao, Z.; Li, X. Highly stereoselective synthesis of 2,3-dihydrofurans via a cascade michael addition-alkylation process: A nitro group as the leaving group. *Chem. Commun.* **2019**, *55*, 6285–6288. [CrossRef]
8. Noble, A.; Anderson, J.C. Nitro-mannich reaction. *Chem. Rev.* **2013**, *113*, 2887–2939. [CrossRef]
9. Sukhorukov, A.Y. C-H reactivity of the α-position in nitrones and nitronates. *Adv. Synth. Catal.* **2020**, *362*, 724–754. [CrossRef]
10. Baiazitov, R.; Denmark, S.E. Tandem [4 + 2]/[3 + 2] cycloadditions. In *Methods and Applications of Cycloaddition Reactions in Organic Syntheses*; Nishiwaki, N., Ed.; John Wiley & Sons: Hoboken, NJ, USA, 2014; pp. 471–550.
11. Nishiwaki, N.; Kumegawa, Y.; Iwai, K.; Yokoyama, S. Development of safely handleable synthetic equivalent of cyanonitrile oxide by 1,3-dipolar cycloaddition of nitroacetonitrile. *Chem. Commun.* **2019**, *55*, 7903–7905. [CrossRef]
12. Ballini, R.; Petrini, M. The nitro to carbonyl conversion (nef reaction): New perspectives for a classical transformation. *Adv. Synth. Catal.* **2015**, *357*, 2371–2402. [CrossRef]
13. Iwai, K.; Asahara, H.; Nishiwaki, N. Synthesis of functionalized 3-cyanoisoxazoles using a dianionic reagent. *J. Org. Chem.* **2017**, *82*, 5409–5415. [CrossRef] [PubMed]
14. Kallitsakis, M.G.; Tancini, P.D.; Dixit, M.; Mpourmpakis, G.; Lykakis, I.N. Mechanistic studies on the Michael addition of amines and hydrazines to nitrostyrenes: Nitroalkane elimination via a retro-aza-henry-type process. *J. Org. Chem.* **2018**, *83*, 1176–1184. [CrossRef] [PubMed]
15. Melot, J.M.; Texier-Boullet, F.; Foucaud, A. Alumina supported potassium fluoride promoted reaction of nitroalkanes with electrophilic alkenes. synthesis of 4,5-dihydrofurans and isoxazoline n-oxides. *Tetrahedron* **1988**, *44*, 2215–2224. [CrossRef]
16. Zhu, C.-Y.; Sun, X.-L.; Deng, X.-M.; Zheng, J.-C.; Tang, Y. Synthesis of isoxazoline N-oxides and its application in the formal synthesis of dehydroclausenamide. *Tetrahedron* **2008**, *64*, 5583–5589. [CrossRef]
17. Zhu, C.-Y.; Deng, X.-M.; Sun, X.-L.; Zheng, J.-C.; Tang, Y. Highly enantioselective synthesis of isoxazoline N-oxides. *Chem. Commun.* **2008**, 738–740. [CrossRef]
18. Chernysheva, N.B.; Maksimenko, A.S.; Andreyanov, F.A.; Kislyi, V.P.; Strelenko, Y.A.; Khrustalev, V.N.; Victor, V. Synthesis of 3,4-diaryl-5-carboxy-4,5-dihydroisoxazole 2-oxides as valuable synthons for anticancer molecules. *Tetrahedron* **2017**, *73*, 6728–6735. [CrossRef]
19. Le Menn, J.C.; Sarrazin, J.; Tallec, A. Formation of isoxazoline N-oxides and dihydrofurans by cyclocondensation of bromo- or chloromalonate carbanion with michael acceptors. *Bull. Soc. Chim. Fr.* **1991**, 562–565.
20. Shi, Z.; Tan, B.; Leong, W.; Wen, Y.; Zeng, X.; Lu, M.; Zhong, G. Catalytic asymmetric formal [4 + 1] annulation leading to optically active cis-isoxazoline N-oxides. *Org. Lett.* **2010**, *12*, 5402–5405. [CrossRef]
21. Chen, X.; Peng, Y.; Yu, W.; Zhang, X.; Shao, X.; Xu, X.; Li, Z. Condition-based selective synthesis of 3,4,5-trisubstituted isoxazoline N-oxides, 4,5-dihydroisoxazoles and isoxazoles. *ChemistrySelect* **2018**, *3*, 6344–6348. [CrossRef]
22. Rouf, A.; Sahin, E.; Tanyeli, C. Divergent synthesis of polysubstituted isoxazoles, isoxazoline N-oxides, and dihydroisoxazoles by a one-pot cascade reaction. *Tetrahedron* **2017**, *73*, 331–337. [CrossRef]

23. Zhang, Y.-R.; Luo, F.; Huang, X.-J.; Xie, J.-W. Water-compatible cascade reaction: An efficient route to substituted 2,3-dihydrofurans. *Chem. Lett.* **2012**, *41*, 777–779. [CrossRef]
24. Chuang, C.-P.; Chen, K.-P.; Hsu, Y.-L.; Tsai, A.-I.; Liu, S.-T. α-nitro carbonyl compounds in the synthesis of 2,3-dihydrofurans. *Tetrahedron* **2008**, *64*, 7511–7516. [CrossRef]
25. Berestovitskaya, V.M.; Baichurin, R.I.; Baichuria, L.V.; Fel'gendler, A.V.; Aboskalova, N.I. Geminally activated β-nitrostyrenes in reactions with cyclic β-diketones. *Russ. J. Gen. Chem.* **2013**, *83*, 1755–1763. [CrossRef]

Sample Availability: Not available.

© 2020 by the authors. Licensee MDPI, Basel, Switzerland. This article is an open access article distributed under the terms and conditions of the Creative Commons Attribution (CC BY) license (http://creativecommons.org/licenses/by/4.0/).

Article

Phenacylation of 6-Methyl-Beta-Nitropyridin-2-Ones and Further Heterocyclization of Products

Eugene V. Babaev *[ID] and Victor B. Rybakov [ID]

Chemistry department, Moscow State University, Leninskie gory, 1 str.3, Moscow 119899, Russia; rybakov20021@yandex.ru
* Correspondence: babaev@org.chem.msu.ru; Tel.: +7-985-997-94-75

Academic Editor: Nagatoshi Nishiwaki
Received: 2 March 2020; Accepted: 4 April 2020; Published: 7 April 2020

Abstract: Reaction between the derivatives of 6-methyl-beta-nitropyridin-2-one and phenacyl bromides was studied, and the yields observed were extremely low. The pyridones were converted via chloropyridines to methoxyderivatives, which were N-phenacylated. N-Phenacyl derivatives of 4,6-dimethyl-5-nitropyridin-2-one under the action of base gave 5-hydroxy-8-nitroindolizine and under acidic conditions gave 5-methyl-6-nitrooxazole[3,2-a]pyridinium salt, which underwent recycization with MeONa to 5-methoxy-8-nitroindolizine.

Keywords: Phenacylation of beta-nitropyridin-2-ones; 8-nitro-5-RO-indolizines; oxazole-pyrrole ring transformation

1. Introduction

Indolizine is an important member of the class of heterocyclic compounds, and many alkaloids have in their structures a saturated or aromatic indolizine moiety. While the chemistry of indolizines has been widely investigated [1], the chemistry of 5-substituted indolizines (**A**, Scheme 1) remains very poor because there are only a few reliable ways for their synthesis. Thus, there are examples of electrophilic substitution of indolizines lithiated at C-5 [2–4] and S_NH reaction of 8-nitroindolizines at this position [5]. The standard Tchitchibabin reaction requires interaction of 6-substituted 2-picolines and alpha-bromoketones; by steric reasons, however, this reaction is almost impossible for 5-X-indolizines. Exceptional case is the cyclization of pyridones **B** (Scheme 1) bearing acceptor (EWG) group at beta-position [6,7]. The last methodology developed in our laboratory is the recyclization of 5-methyl substituted oxazolo[3,2-a]pyridinium salts **C** via pyridinium betaine **D** (Scheme 1), which leads to 5-substituted indolizines [8,9]. In turn, the salts **C** are available via acidic cyclization of pyridones **B** [10]:

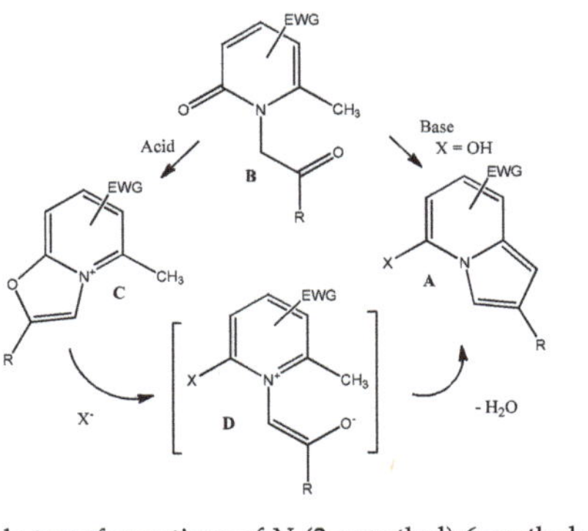

Scheme 1. Possible transformations of N-(2-oxoethyl)-6-methylpyridin-2-ones.

The aim of this work was to test the validity of Scheme 1 for the case of more powerful acceptor a nitro group (EWG = NO_2) by choosing the 6-methyl-beta-nitropyridin-2-ones as the parent compounds. Obtained by such strategy 6- or 8-nitroindolizines with the substituent X at position 5 may undergo easy nucleophilic substitution at C-5. In addition to the route from **B** to **A**, one could consider one more alternative strategy from **C** to **A**.

2. Results and Discussion

2.1. Synthesis of Isomers and Homologues of Beta-Nitro-6-Methylpyridin-2-One (1a–d)

The simplest scheme for the synthesis of 3- and 5-nitro derivatives of 6-methylpyridin-2-one was the diazotation of 2-amino-3(5)-nitro-6-picolines, which, in turn, are available by nitration of commercial 2-amino-6-picoline to 2-nitramino-6-picoline (95%) and its further acidic rearrangement (51%) and steam distillation (leading to 39% of 5-nitro- and 15% of 3-nitropyridin-2-ones). By this way (Scheme 2) both isomers, namely 6-methyl-3-nitropyridin-2-one (**1a**, yield 32% [11]) and 6-methyl-5-nitropyridin-2-one (**1b**, yield 25% [12]), were obtained; the low yield at the last step may be due to high solubility of products in acetone for recrystallization.

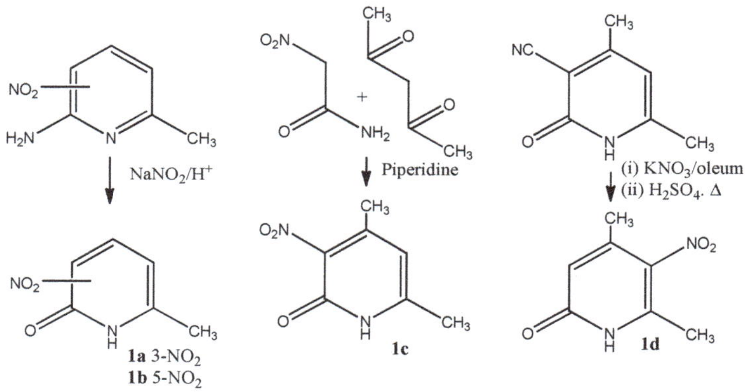

Scheme 2. Synthesis of the parent homologues of beta-nitro-6-pyridin-2-ones.

In addition, two 4-methyl homologues of pyridones **1a,b** were described in literature. To prepare 4,6-dimethyl-3-nitropyridin-2-one (**1c**) we performed the cyclization of acetylacetone with nitroacetamide and obtained the target compound **1c** with the yield 37% [13]. For preparation of nitroacetamide we used the isonitroacetoacetic ester, which was dangerous for synthetic chemists due to ability to detonate above 100 °C. (In addition, freshly isolated dry nitroacetamide was capable to self-ignition upon contact with air.) Finally, the synthesis of 4,6-dimethyl-5-nitropyridin-2-one (**1d**, yield 19% [14]) was achieved in two steps by nitration of Guaresci pyridine (obtained with the yield 93% and nitrated with the yield 47%) and consequent removal of the CN group in diluted sulfuric acid (low yield at this step may be due to solubility of product both in acid and in alkali used for neutralization).

2.2. Attempts of Direct Phenacylation of Homologues of Beta-Nitropyridin-2-Ones

Alkylation (and phenacylation) of pyridin-2-ones may occur either at *N*- or *O*-atoms [9]. Based on the literature data of reaction of phenacyl bromides with 4,6-dimethylpyridin-2-ones (bearing electron withdrawing group CN, $CONH_2$, CO_2R at C-5) [6,7,15–18]) we expected that alkylation of sodium salts of 6-methyl-beta-nitropyridin-2-ones (**1a–d**) would also proceed at the nitrogen atom. As an additional argument, Na-salt of 5-nitropyridin-2-one also underwent *N*-phenacylation [19]. We believed that the resulting mixtures of *N*- and *O*-phenacyl derivatives from **1a–d** could be chromatographically separated, since it is often mentioned in the literature that the chromatographic behavior of N- and O-isomers is different.

It turned out, however, that when trying to phenacylate the Na-salt of nitropyridones **1a–d**, the reactions proceeded with an extremely low yield. As can be seen, a combination of factors (a sterically hindered nitrogen atom deactivated by a nitro group) prevents N-alkylation. The result obtained completely excluded the opportunity to implement our planned strategy. In search of a possible solution to the problem, we drew attention to the other strategy described in the literature for the regioselective synthesis of pyridones with a phenacyl residue at the nitrogen atom using 2-methoxypyridines [9,20,21]. In this case, O-demethylation in the intermediate salt was apparently due to the attack of a rather weak nucleophile, the bromide ion. As a result, the methyl group acted as a protective group, and this method allowed selective and reliable preparation of N-phenacyl derivatives. For our purposes, we should use 2-methoxy-beta-nitropyridines (with a kind of protection—a methoxy group, which not only prevents competitive O-alkylation but also acts as a donor residue that promotes N-alkylation).

2.3. Synthesis of Beta-Nitro-2-Methoxypyridine Homologues and Their Phenacylation

An analysis of the literature showed that it is not a serious problem to convert the beta-nitropyridin-2-ones into 2-methoxypyridines with the intermediacy of 2-chloropyridines. The conversion of pyridin-2-ones to 2-chloropyridines proceeded with high yields upon boiling with $POCl_3$ (Scheme 3), and the melting points of the obtained chloropyridines **2a–d** coincided with the published data (Table 1). The next stage—the replacement of Cl with a MeO group—was carried out by boiling with MeONa in methanol. The yields and melting points of the obtained substances **3a–d** are shown in Table 1.

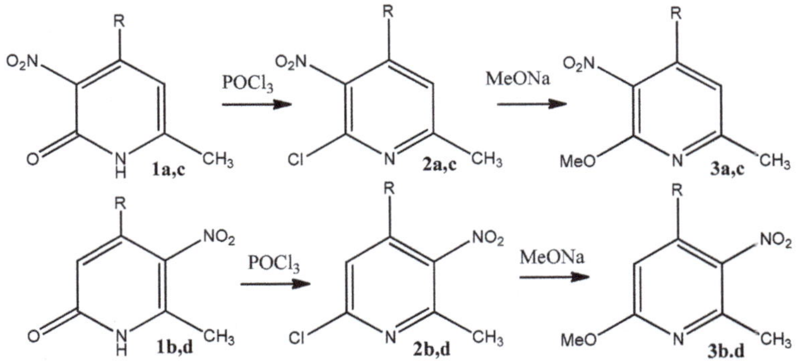

Scheme 3. Conversion of beta-nitropyridin-2-ones to 2-cloro- and 2-methoxy derivatives. **1–3**: **a, b** R = H; **c, d** R = CH_3.

Table 1. Yields, mp and literature references for compounds **2a–d** and **3a–d**.

No	Yield, %	Mp	Mp, Lit	Reference
2a	45	68	70–70.8	[22]
			67–69	[23]
3a	32	57	57–58	[24]
2b	25	54–55	54	[25]
3b	36	64	64–65	[26]
2c	64	47	47–48	[27]
3c	90	104–105	104	[28]
2d	85	54	54–55	[27]
3d	92	59–60	60	[14]

After numerous attempts to phenacylate 2-methoxypyridines **3**, we found the only acceptable method—melting of the starting reagents, Scheme 4. Such melting of reagents (in comparison with their boiling in acetonitrile) increased the yields 17 times, and they achieved 35%. The structure of *N*-phenacyl derivatives clearly followed from the spectral data. The ^{1}H-NMR spectra of phenacylpyridones **4a,b** contain the expected signals of the phenacyl residue and a fragment of nitropyridone. The final confirmation of the structure of the *N*-phenacylpyridone **4a** was obtained by X-ray diffraction, Figure 1. The yields for other methoxypyridines (**3a–c**) were much lower.

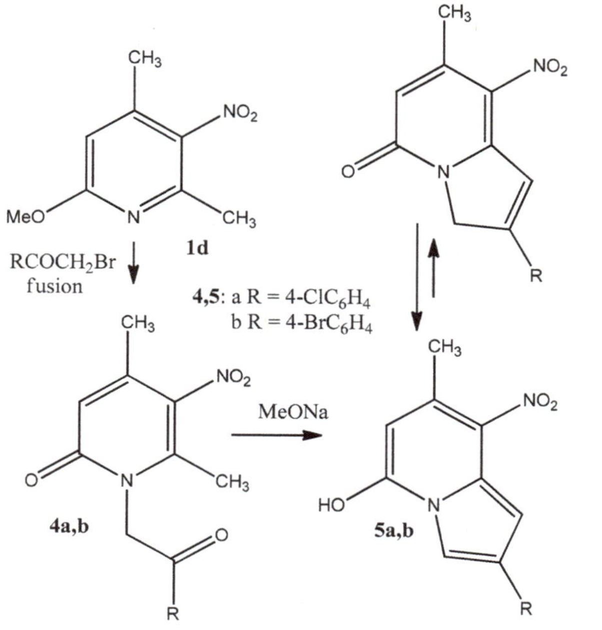

Scheme 4. Phenacylation of nitropyridones and consequent cyclization.

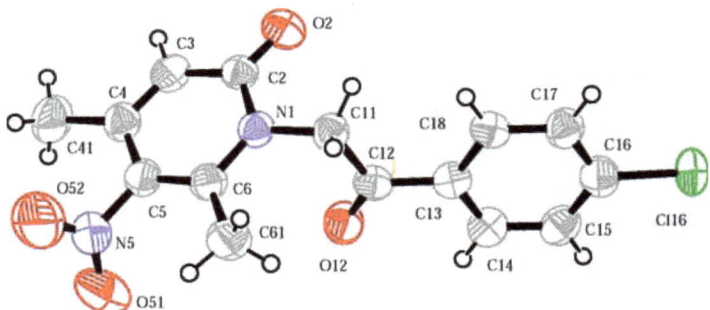

Figure 1. The general view of molecular structure of **4a** in representation of atoms by displacement parameters ellipsoids (p = 30%).

2.4. Cyclocondensation of N-Phenacylpyridones under Basic Condition

For studies of the cyclization of phenacylpyridones, we used compounds **4a,b** obtained in acceptable quantities (see Experimental section). In the solution of **4a,b** in MeOH, an intense dark red color is observed. The neutralization of the reaction mixture allowed us to identify individual powdery compounds of dark red color. Their solutions turned black-green. The data of ^{1}H- and ^{13}C-NMR spectra showed that both compounds are substituted indolizines **5a,b** with the structure of 5-hydroxy tautomers, Scheme 4. This observation contradicted to the structure of 6-CN-5-hydroxy indolizines to which 5-oxo-3-CH$_2$-type of tautomers was assigned [6,7]. In the ^{1}H-NMR spectrum, we observed three singlets (in addition to multiplet of aryl residue and CH$_3$ singlet). Two of them were from protons H-1 and H-3 of the pyrrole fragment, and the third one—at 4.4–5.4 ppm—was a singlet of the proton H-6 of the pyridine fragment. The last peak was shifted to a high field due to the ortho-located hydroxy group, whose signal appeared as a broadened peak in the region of 3.1–3.4 ppm. The confirmation that in this case indolizine existed precisely in the hydroxy form was the appearance in the IR spectrum of the characteristic vibration frequency of the OH group at 3437 cm^{-1}.

The reasons why 5-hydroxy (oxo) indolizines existed in the oxy or oxo form depended, obviously, on the nature of the additional acceptor substituent in the pyridine nucleus. In the case of 6-cyano derivatives, the tautomeric equilibrium was completely shifted towards the oxo form, while in the case of 8-nitroindolizins, the oxy form prevailed. These groups were likely to have different effects on the acidity of the 5-OH group and the basicity of the C-3 atom of the pyrrole moiety (onto which the proton of the hydroxyl group can migrate).

By analogy with 6-cyanoindolizinones [7], we expected that the action of phosphorus oxychloride on the corresponding nitroindolizinoles would lead to the replacement of 5-oxo/oxy-group with chlorine. It turned out, however, that when both hydroxyindolizines were heated with POCl$_3$, complete resinification was observed. Probably the phenolic nature of the obtained indolizines somehow prevented the occurrence of such a transformation.

2.5. Synthesis of 5-Substituted Indolizine via Oxazolopyridinium Salt

As we have seen, it was possible to obtain intermediate compounds, representatives of a previously unknown class of 5-hydroxyindolizines, and their structure was very interesting. It turned out, however, that these compounds were unpromising for introduction of any other functions to position 5 (since they cannot be converted into 5-chloro derivatives). This can mean that it was completely impossible to vary the residue at position 5 in the series of 6(8)-nitroindolizines within the framework of our chosen strategy.

We were able to show that the precursors of 5-hydroxyindolizines, phenacylpyridones **4**, were very promising synthons for producing 6 (8)-nitroindolizines with a substituent other than the hydroxy group in position 5. It is known that such phenacylpyridones can close two different cycles under the action of bases and acids. In the last case cyclodehydration may lead to closure of the oxazolium cycle. We found that, under the action of concentrated sulfuric acid, phenacylpyridone **4a** were smoothly cyclized to the corresponding oxazolopyridinium cation **6**, isolated in the form of perchlorate, Scheme 5. In the ^{1}H-NMR spectrum of the obtained salt, a singlet in low field appeared at 9.72 ppm, which corresponded to the proton of the newly formed oxazolium ring. The final proof of the structure of the obtained heterocycle **6** was obtained by X-ray diffraction, Figure 2a.

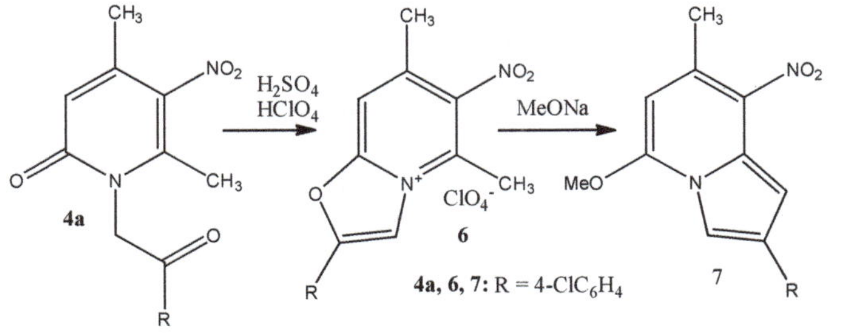

Scheme 5. Closure of oxazolium ring in N-phenacylpyridone and further recyclization.

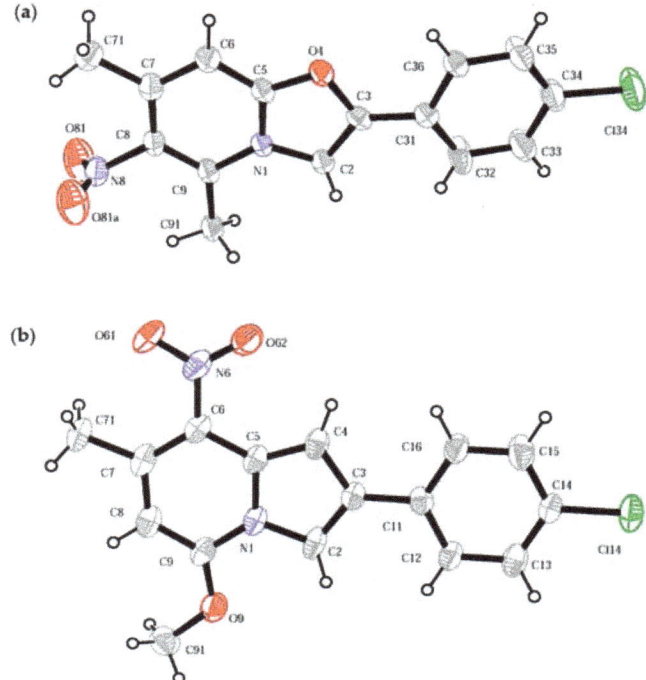

Figure 2. The general view of molecular structures of **6** (**a**) and **7** (**b**) in representation of atoms by displacement parameters ellipsoids (p = 30%). A perchlorate anion for **6** and solvate acetone molecule for **7** are omitted for clarity.

The resulting compound **6** was capable of easily reacting with nucleophiles. It turned out that in the reaction with MeONa, compound **6** underwent a rearrangement with the destruction of the oxazolium ring and the closure of a new pyrrole ring, forming 5-methoxy-8-nitroindolizine **7**, Scheme 5. The structure of the obtained compound **7** was unambiguously proved by the X-ray diffraction method, Figure 2b. (In the lattice there is a solvent molecule—acetone.)

Interestingly, the obtained 5-methoxy-8-nitroindolizine **7** was isostructural to the 5-hydroxy-8-nitroindolizines **5a,b** obtained above. Comparing the electronic absorption spectra of these two isostructural (and pi-isoelectronic) systems, one can conclude that the structure of the bands was similar.

3. Materials and Methods

3.1. General Information

^{1}H-NMR spectra were recorded on a Bruker AC 400 instrument (Bruker, Bremen, Germany, operating frequency 360 MHz,), ^{13}C-NMR spectra were recorded on a frequency of 100 MHz. Chemical shifts are measured on a δ-scale and are given in parts per million; *J* given in Hertz. The reaction progress was monitored by TLC on Silufol UV-254 plates (Merck KGaA, Darmstadt, Germany), and the TLC manifestation was carried out by UV radiation (wavelengths of 254 and 365 nm), iodine vapor, oxidation in a sulfuric aqueous solution of KMnO$_4$, Ehrlich reagent or ninhydrin. Chromatographic separation was performed on columns or glass plates using silica gel with a particle size of 40–60 µm (Merck, KGaA, Darmstadt, Germany). Reagents from Acros (Fisher Scientific, Leicestershire, UK), Merck (Merck KGaA, Darmstadt, Germany) and Aldrich (Sigma-Aldrich Company

Ltd., Dorset, UK) were used as starting materials for the syntheses. They were introduced into the reactions without additional purification. Elementary analysis data for new compounds are given in Table 2.

Table 2. Elementary analysis data for all new compounds.

Compound (No.)	Formula	Calculated			Found		
		C	H	N	C	H	N
N-(p-chlorophenacyl)-4,6-dimethyl-5-nitropyridin-2-one (4a)	$C_{15}H_{13}ClN_2O_4$	56.17	4.09	8.73	55.82	4.13	8.68
N-(p-bromophenacyl)-4,6-dimethyl-5-nitropyridin-2-one (4b)	$C_{15}H_{13}BrN_2O_4$	49.34	3.59	7.67	49.13	3.62	7.64
2-p-chlorophenyl-5-hydroxy-7-methyl-8-nitroindolizine (5a)	$C_{15}H_{11}ClN_2O_3$	59.52	3.66	9.25	59.13	3.71	9.19
2-p-Bromophenyl-7-methyl-8-nitro-5-hydroxyindolizine (5b)	$C_{15}H_{11}BrN_2O_3$	51.90	3.19	8.07	51.51	3.25	8.01
2-(p-Chlorophenyl)-5,7-dimethyl-6-nitrooxazolo [3,2-a]pyridinium perchlorate (6)	$C_{15}H_{12}ClN_2O_3 \cdot ClO_4$	44.69	3.00	6.95	44.40	3.05	6.90
2-(p-Chlorophenyl)-5-methoxy-7-methyl-8-nitroindolizine (7)	$C_{16}H_{13}ClN_2O_3$	60.67	4.14	8.84	60.29	4.18	8.79

* means ionic compound.

3.2. Synthesis

6-Methyl-3- and 5-nitropyridin-2-ones (**1a,b**) were obtained by diazotation of 2-amino-3(5)-nitro-6-picolines [11,12]. 4,6-Dimethyl-3-nitropyridin-2-one (**1c**) obtained by cyclization of acetylacetone with nitroacetamide [13]. 4,6-Dimethyl-5-nitropyridin-2-one (**1d**) was obtained in two steps by nitration of Guaresci pyridine and further hydrolysis [14].

Studying of phenacylation of sodium salt 4,6-dimethyl-5-nitropyridin-2-one (**1d**). In methanol. To a solution of 0.483 g (0.021 mol) of sodium metal in 100 mL of absolute methanol, 3.36 g (0.02 mol) of nitropyridone **1d** and 4.9 g (0.021 mol) of p-chlorophenacyl bromide are added with vigorous stirring. The mixture was stirred for 12 h at 55–60 °C and controlled by TLC. Only traces of product could be observed. In benzene/xylene. To a benzene suspension of 0.5 g (2.9 mmol) sodium salt of pyridone **1d** 0.8 g (35 mmol) of p-bromophenacyl bromide was added. After 60 h at room temperature TLC monitoring showed the presence of only starting materials. The solvent was evaporated and the mixture was dissolved in o-xylene and boiled for another 6 h. TLC control showed the presence of only traces of products.

Conversion of pyridin-2-ones (**1a–d**) to 2-chloropyridines (**2a–d**). The mixture of 0.036 mol of nitropyridone-2 (**1a–d**), 4 mL of POCl$_3$ and 0.036 mol of PCl$_5$ was maintained at 120 °C for 4 h. The mixture was left at 120 °C in a Wood alloy for 1.5 h. The mixture was cooled to RT, poured into excess of ice water and the brown precipitate was filtered off, dried and recrystallized from hexane. The yields, m.p. of products **2a–d** and literature references are given in Table 2.

Conversion of 2-chloropyridines (**2a–d**) to 2-methoxypyridines (**3a–d**). To a solution of MeONa (obtained by dissolving 0.388 g (0.017 mol) of sodium metal in 15 mL of absolute MeOH) 0.016 mol 2-chloropyridine **2a–d** was added. The mixture was boiled for 4 h, the precipitated NaCl was filtered off, the filtrate was evaporated and the residue was chromatographed on a column (SiO$_2$, chloroform). The yields, m.p. of products **3a–d** and literature references are given in Table 1.

N-(p-chlorophenacyl)-4,6-dimethyl-5-nitropyridin-2-one (**4a**). A mixture of 1.3 g (0.007 mol) of methoxypyridine **3e** and 1.75 g (0.007 mol) of p-chlorophenacyl bromide was dissolved in 10 mL of acetonitrile and the solution was boiled for 30 h. The solvent was evaporated in vacuo. The resulting mixture was kept for 17 h at 100 °C and 15 h at 120 °C, cooled to room temperature and chromatographed on a column (SiO$_2$, chloroform). Product **4a** (0.14 g) was obtained. Yield 6%; m.p. 190–191 °C. ^{1}H-NMR Spectrum (DMSO-d_6): 8.10 (2H, m, Ar); 7.57 (2H, m, Ar); 6.30 (1H, s, 3-H); 5.65 (2H, s, CH$_2$); 2.30 (3H, s, 6-CH$_3$); 2.22 (3H, s, 4-CH$_3$). The molecular structure is shown in Figure 1 [28].

N-(p-bromophenacyl)-4,6-dimethyl-5-nitropyridin-2-one (**4b**). Obtained in a similar manner from methoxypyridine **3d** and p-bromophenacyl bromide. Yield 8%, m.p. 192 °C. ^{1}H-NMR Spectrum

(CDCl$_3$): 7.92 (2H, m, Ar); 7.78 (2H, m, Ar); 6.94 (1H, s, 3-H); 5.74 (2H, s, CH$_2$); 2.30 (3H, s, 6(4)-CH$_3$); 2.25 (3H, s, 4(6)-CH$_3$).

Phenacylation of methoxypyridine **3d**: synthesis optimization without use of a solvent.

(1) The reaction of **3d** with p-chlorophenacyl bromides was carried out in a sealed glass ampoule, the reaction mass was heated in an oven for 25 h at temperature of 120–150 °C. TLC analysis in pure chloroform showed the presence of the desired products. After chromatography, 0.378 g of a pure yellow substance, identical in TLC and m.p. with the N-isomer **4a**. Yield ~10%.

(2) The substances were placed in a 50 mL flask equipped with a reflux condenser. The mixture was kept at 120 °C for 7 h in a Wood alloy. After 5 h of boiling, the yellow liquid began to accumulate on the walls of the flask and flowed down. TLC analysis showed that it is phenacyl bromide. Then, after the cessation of gas formation, the temperature was raised to 200 °C (continued gas formation), and the mixture was kept at this temperature for another 3 h. The reaction mixture was applied to silica gel and chromatographed with CHCl$_3$, then CHCl$_3$—EtOH. Recrystallized from acetone. The yield of N-isomer **4a** was 15%.

(3) 4 g of methoxypyridine **3d** and a twofold excess of phenacyl bromide were taken. The substances are placed in a 50 mL flask equipped with a reflux condenser. The mixture was heated in a Wood alloy at 125 °C for 20 h. After this, the mixture was poured into a large amount of boiling petroleum ether and the precipitate was filtered off. A petroleum extract containing pure starting materials was reacted back. The precipitate was purified from the crude oil by chromatography, eluting with chloroform. The yield of the target product **4a** was 35%.

Synthesis of 5-hydroxy-8-nitroindolizines **5a,b** (General methodology). A total of 0.6 g of phenacylpyridone **4a** (1.6 mmol) was dissolved in 100 mL of MeOH. The calculated amount of Na (38 mg) was dissolved in 50 mL of MeOH. A solution of MeONa was added with stirring to a solution of phenacylpyridone. The solution turned raspberry colored. In 30 min after the start of the reaction the calculated amount of HOAc was added to neutralize MeONa. The reaction mixture was diluted with diethyl ether, and the product and NaOAc precipitated. The solvent was distilled off on a rotary evaporator, the mixture was dissolved in a minimum amount of acetone and filtered from NaOAc. The solution was evaporated giving a brick-red substance, 2-p-chlorophenyl-5-hydroxy-7-methyl-8-nitroindolizine (**5a**). Yield 92%. m.p. > 250 °C (decomp.). ^{1}H-NMR Spectrum (CD$_3$OD): 7.80 (1H, m, 3-H); 7.72 (2H, m, Ar); 7.47 (2H, m, Ar); 7.29 (1H, m, 1-H); 5.37 (1H, s, 6-H); 3.40 (1H, br s, OH); 2.57 (3H, s, CH$_3$).

2-p-Bromophenyl-7-methyl-8-nitro-5-hydroxyindolizine (**5b**). Yield 90%, m.p. > 250 °C (decomp.). ^{1}H-NMR Spectrum (CDCl$_3$): 7.34 (1H, m, 3-H); 7.17 (2H, m, Ar); 7.08 (2H, m, Ar); 6.82 (1H, m, 1-H); 4.37 (1H, s, 6-H); 3.10 (1H, br s, OH); 2.13 (3H, s, CH$_3$).

2-(p-Chlorophenyl)-5,7-dimethyl-6-nitrooxazolo [3,2-*a*]pyridinium perchlorate (**6**). A mixture of 0.1 g (0.3 mmol) of phenacylpyridone **5a** and 1 mL of concentrated sulfuric acid was maintained at 22 °C for 18 h. Then, 0.2 mL of 70% HClO$_4$ was added to the mixture, incubated for 1 h, poured into 100 mL of absolute ether and the precipitate formed was filtered off. The yield of perchlorate is 92%; m.p. 295–297 °C (decomp.). ^{1}H-NMR Spectrum (DMSO-d_6): 9.72 (1H, s, 3-H); 8.60 (1H, s, 8-H); 8.07 (2H, m, Ar); 7.82 (2H, m, Ar); 2.89 (3H, s, 5-CH$_3$); 2.67 (3H, s, 7-CH$_3$). ^{13}C-NMR (DMSO-d_6): 15.4; 18.5; 110.6; 112.1; 122.6; 127.5; 130.1; 137.1; 137.8; 144.8; 145.8; 152.2; 152.4. X-Ray data see Figure 2a. The molecular structure is seen in Figure 2a (perchlorate anion is omitted for clarity) [29].

2-(p-Chlorophenyl)-5-methoxy-7-methyl-8-nitroindolizine (**7**). A total of 200 mg (0.49 mmol) of oxazolopyridinium salt **6** was added to a solution of 10 mg of sodium in 10 mL of methanol. The mixture was kept for 1 day at 22 °C and the precipitate formed was filtered off. Product **7** (110 mg) was obtained. Yield 73%, m.p. 175–176 °C. ^{1}H-NMR Spectrum: 7.50 (6H, m, Ar); 5.73 (1H, s, 6-H); 4.22 (3H, s, O-CH$_3$); 2.69 (3H, s, CH$_3$). The molecular structure see Figure 2b (solvate acetone molecule is omitted for clarity) [30].

3.3. X-ray Diffraction Studies

For single crystals of compounds **4a** and **6**, the experimental intensities of diffraction reflections were obtained on a CAD-4 diffractometer (λ Cu Kα radiation, graphite monochromator, ω scanning at room temperature, Enraf-Nonius, Delft, Netherlands), for compound **7** on a CAD-4 diffractometer (λ Mo Kα radiation, graphite monochromator, ω-scan, at room temperature). All subsequent calculations were performed as part of the SHELX software package [31]. The crystallographic data for the studied structures were deposited in the Cambridge Structural Database with the numbers CCDC 1986879 for **4a**, CCDC 1986629 for **6**, CCDC 1984804 for **7**.

4. Conclusions

In conclusion, the reaction between the derivatives of 6-methyl-beta-nitropyridin-2-one (**1a–d**) and phenacyl bromides did not occur for steric and electronic reasons. However, 4,6-dimethyl-2-methoxy-5-nitropyridine **3d** on fusion with bromoketones could be *N*-phenacylated. Its *N*-phenacyl derivatives **4a,b** under the action of base gave 5-hydroxy-8-nitroindolizines (**5a,b**) and under acidic conditions gave 5-methyl-6-nitrooxazolo[3,2-*a*]pyridinium salt **6**, which underwent recycization with MeONa to 5-methoxy-8-nitroindolizine **7**.

Author Contributions: E.V.B. formulated the goals, managed performance of all experimental work and wrote the paper. V.B.R. performed X-ray analysis. All authors have read and agreed to the published version of the manuscript.

Funding: This work was supported by RFBR grant No19-53-53009 GFEN_a.

Acknowledgments: We thank R. Lukov for help in experimental work. The authors are grateful to Thermo Fisher Scientific Inc., Analitika (Moscow), and personally to A.A. Makarov for providing the mass spectrometric equipment used in this work.

Conflicts of Interest: The authors declare no conflicts of interest.

References

1. Flitsch, W. Pyrroles with fused six-membered heterocyclic rings: A-fused. In *Comprehensive Heterocyclic Chemistry*; Katritzky, A., Rees, C.W., Eds.; Pergamon Press: Oxford, UK, 1984; Volume 4, pp. 443–496.
2. Renard, M.; Gubin, J. Metallation of 2-Phenylindolizine. *Tetrahedron Lett.* **1992**, *33*, 4433–4434. [CrossRef]
3. Kuznetsov, A.G.; Bush, A.A.; Rybakov, V.B.; Babaev, E.V. An Improved Synthesis of Some 5-Substituted Indolizines Using Regiospecific Lithiation. *Molecules* **2005**, *10*, 1074–1083. [CrossRef] [PubMed]
4. Rzhevskii, S.A.; Rybakov, V.B.; Khrustalev, V.N.; Babaev, E.V. Reactions of 5-Indolizyl Lithium Compounds with Some Bielectrophiles. *Molecules* **2017**, *22*, 661. [CrossRef] [PubMed]
5. Kost, A.N.; Sagitullin, R.S.; Gromov, S.P. Nucleophilic amination and recyclization of the indolizine nucleus. *Heterocycles* **1977**, *7*, 997–1001.
6. Gevald, K.; Jansch, H.J. 3-Amino-furo[2.3-b]pyridine. *J. Prakt. Chem.* **1976**, *318*, 313–320.
7. Babaev, E.V.; Vasilevich, N.I.; Ivushkina, A.S. Efficient synthesis of 5-substituted 2-aryl-6-cyanoindolizines via nucleophilic substitution reactions. *Beilstein J. Org. Chem.* **2005**, *1*, 9–11. [CrossRef]
8. Babaev, E.V. Fused Munchnones in Recyclization Tandems. *J. Heterocycl. Chem.* **2000**, *37*, 519–526. [CrossRef]
9. Babaev, E.V.; Alifanov, V.L.; Efimov, A.V. Oxazolo[3,2-a]pyridinium and Oxazolo[3,2-a]pyrimidinium Salts in Organic Synthesis. *Russ. Chemical Bull.* **2008**, *57*, 845–862. [CrossRef]
10. Bradsher, C.K.; Zinn, M.F. Oxazolo[3,2-a] pyridinium salts. *J. Heterocycl. Chem.* **1967**, *4*, 66–70. [CrossRef]
11. Krasnaya, Z.A.; Stytsenko, T.S.; Prokof'ev, E.P.; Yakovlev, I.P.; Kucherov, V.F. Reaction of enaminocarbonyl compounds with nitroacetic ester. *Bull. Acad. Sci. USSR* **1974**, *23*, 809–816. [CrossRef]
12. Goerlitzer, K.; Wilpert, C.; Ruebsamen-Waigmann, H.; Suhartono, H.; Wang, L.; Immelmann, A. Pyrido[3,2-e][1,4] diazepine - Synthese und Prüfung auf Anti-HIV-1-Wirkung. *Arch. Pharm.* **1995**, *328*, 247–256. [CrossRef] [PubMed]
13. Kislyi, V.P.; Shestopalov, A.M.; Kagramanov, N.D.; Semenov, V.V. Synthesis of 3-nitropyrid-2(1H)-ones from C-nitroacetamide and 1,3-dicarbonyl compounds. *Russ. Chem. Bull.* **1997**, *46*, 539–542. [CrossRef]

14. Mariella, R.P.; Callahan, J.J.; Jibril, A.O. Some novel color reactions of some pyridine derivatives. *J. Org. Chem.* **1955**, *20*, 1721–1728. [CrossRef]
15. Rybakov, V.B.; Babaev, E.V.; Paronikyan, E.G. X-Ray Mapping in Heterocyclic Design: 18. X-Ray Diffraction Study of a Series of Derivatives of 3-Cyanopyridine-2-one with Annelated Heptane and Octane Cycles. *Crystallogr. Rep.* **2017**, *62*, 219–231. [CrossRef]
16. Rybakov, V.B.; Babaev, E.V. Transformations of Substituted Oxazolo[3,2-a]Pyridines to 5,6-Disubstituted Indolizines: Synthesis And X-ray Structural Mapping. *Chem. Heterocycl. Comp.* **2014**, *50*, 225–236. [CrossRef]
17. Okul', E.M.; Rybakov, V.B.; Babaev, E.V. The structure of products of phenacylation and subsequent (re)cyclizations of 3-acetyl-4,6-dimethylpyridin-2(1H)-one according to X-ray structural analysis. *Chem. Heterocycl. Comp.* **2017**, *53*, 997–1002. [CrossRef]
18. Feklicheva, E.M.; Rybakov, V.B.; Babaev, E.V.; Oficerov, E.N. Physical-chemical investigations of transformations in the series of 2,4-dimethyl-6-oxo-1,6-dihydropyridine-3-carboxanmide. Part 1. Synthesis and X-ray study of derivatives of 2,4-dimethyl-6-oxo-1,6-dihydropyridine-3-carboxanmide. *Butlerov Communications* **2019**, *60*, 1–23.
19. Bush, A.A.; Babaev, E.V. Synthesis of 6-Nitroderivatives of Oxazolo[3,2-a]pyridines and their Reactions with Nucleophiles. *Molecules* **2003**, *8*, 460–466. [CrossRef]
20. Babaev, E.V.; Efimov, A.V.; Maiboroda, D.A. Hetarenes with a nitrogen bridging atom. 1. Phenacylation of 2-substituted 6-methylpyridines. *Chem. Heterocycl. Comp.* **1995**, *31*, 962–968. [CrossRef]
21. Babaev, E.V.; Tsisevich, A.A.; Al'bov, D.V.; Rybakov, V.B.; Aslanov, L.A. Heterocycles with a bridgehead nitrogen atom. Part 16. Assembly of a peri-fused system from an angular tricycle by recyclization of an oxazole ring into pyrrole one. *Russ. Chem. Bull.* **2005**, *54*, 259–261. [CrossRef]
22. Parker, E.D.; Shive, W. Substituted 2-Picolines Derived from 6-Amino-2-picoline. *J. Amer. Chem. Soc.* **1947**, *69*, 63–67. [CrossRef]
23. Remennikov, G.Y.; Kurilenko, L.K.; Boldyrev, I.V.; Cherkasov, V.M. The recyclization of 5-nitropyrimidine and its methoxy derivatives upon reaction with the acetylacetone carbanion. *Chem. Heterocycl. Comp.* **1987**, *23*, 422–425. [CrossRef]
24. Frydman, B.; Reil, S.J.; Boned, J.; Rapoport, H. Synthesis of substituted 4- and 6-azaindoles. *J. Org. Chem.* **1968**, *33*, 3762–3766. [CrossRef]
25. Baumgarten, H.E.; Su, H.C.-F. Synthesis of 3- and 5-nitro-2-picoline and derivatives. *J. Amer. Chem. Soc.* **1952**, *74*, 3828–3830. [CrossRef]
26. Sawanishi, H.; Tajima, K.; Tsuchiya, T. Studies on Diazepines. XXVIII. Syntheses of 5H-1,3-Diazepines and 2H-1,4-Diazepines from 3-Azidopyridines. *Chem. Pharm. Bull.* **1987**, *35*, 4101–4109. [CrossRef]
27. Kislyi, V.P.; Semenov, V.V. Investigation of the regioselectivity of alkylation of 3-nitropyridine-2(1H)-ones. *Russ. Chem. Bull.* **2001**, *50*, 460–463. [CrossRef]
28. Babaev, E.V.; Rybakov, V.B. CCDC 1986879. *Experimental Crystal Structure Determination.* 2020. Available online: https://doi.org/10.5517/ccdc.csd.cc24phw7 (accessed on 27 February 2020).
29. Babaev, E.V.; Rybakov, V.B. CCDC 1986629. *Experimental Crystal Structure Determination.* 2020. Available online: https://doi.org/10.5517/ccdc.csd.cc24p7tx (accessed on 27 February 2020).
30. Babaev, E.V.; Rybakov, V.B. CCDC 1984804. *Experimental Crystal Structure Determination.* 2020. Available online: https://doi.org/10.5517/ccdc.csd.cc24mby2 (accessed on 18 February 2020).
31. Sheldrick, G.M. Crystal structure refinement with SHELX. *Acta Cryst.* **2015**, *C71*, 3–8. [CrossRef]

Sample Availability: Samples of the compounds **6** and **7** are available from the authors.

© 2020 by the authors. Licensee MDPI, Basel, Switzerland. This article is an open access article distributed under the terms and conditions of the Creative Commons Attribution (CC BY) license (http://creativecommons.org/licenses/by/4.0/).

Article

The Cyclic Nitronate Route to Pharmaceutical Molecules: Synthesis of GSK's Potent PDE4 Inhibitor as a Case Study

Evgeny V. Pospelov [1,2], Ivan S. Golovanov [1], Sema L. Ioffe [1] and Alexey Yu. Sukhorukov [1,3,*]

1. N.D. Zelinsky Institute of Organic Chemistry, Russian Academy of Sciences, 119991 Moscow, Russia; evpos00@mail.ru (E.V.P.); cell-25@yandex.ru (I.S.G.); iof@ioc.ac.ru (S.L.I.)
2. Department of Chemistry, M.V. Lomonosov Moscow State University, 119991 Moscow, Russia
3. Department of Innovational Materials and Technologies Chemistry, Plekhanov Russian University of Economics, 117997 Moscow, Russia
* Correspondence: sukhorukov@ioc.ac.ru; Tel.: +7-499-135-53-29

Academic Editor: Nagatoshi Nishiwaki
Received: 3 July 2020; Accepted: 4 August 2020; Published: 8 August 2020

Abstract: An efficient asymmetric synthesis of GlaxoSmithKline's potent PDE4 inhibitor was accomplished in eight steps from a catechol-derived nitroalkene. The key intermediate (3-acyloxymethyl-substituted 1,2-oxazine) was prepared in a straightforward manner by tandem acylation/(3,3)-sigmatropic rearrangement of the corresponding 1,2-oxazine-*N*-oxide. The latter was assembled by a (4 + 2)-cycloaddition between the suitably substituted nitroalkene and vinyl ether. Facile acetal epimerization at the C-6 position in 1,2-oxazine ring was observed in the course of reduction with NaBH$_3$CN in AcOH. Density functional theory (DFT) calculations suggest that the epimerization may proceed through an unusual tricyclic oxazolo(1,2)oxazinium cation formed via double anchimeric assistance from a distant acyloxy group and the nitrogen atom of the 1,2-oxazine ring.

Keywords: C–H functionalization; total synthesis; pyrrolidines; anchimeric assistance; epimerization; PDE4 inhibitors

1. Introduction

Cyclic nitronates (1,2-oxazine-*N*-oxides **1** and isoxazoline-*N*-oxides **2**) are useful intermediates in the synthesis of complex nitrogen containing scaffolds due to their versatile reactivity as 1,3-dipoles and accessibility from nitroalkenes (Scheme 1a) [1–10]. Denmark's group extensively exploited the inter- and intramolecular (3 + 2)-cycloaddition reactions with six-membered cyclic nitronates **1** to construct various bi- and polycyclic nitroso acetal frameworks **3**, which were then converted into fused pyrrolidine derivatives by an intramolecular reductive amination (Scheme 1b) [1,11]. Using this strategy, total syntheses of numerous pyrrolizidine and indolizidine alkaloids [12–14], as well as (5.5.5.5)- and (5.5.5.4)-azafenestanes [15,16], were accomplished.

Our group has a long-term interest in developing another approach towards the modification of cyclic nitronates, which utilizes C–H functionalization of the position next to the nitronate group (α-C-atom, Scheme 1c) [17]. Some time ago, we demonstrated that upon silylation, cyclic nitronates **1** and **2** are transformed into *N*-siloxyenamines **4**, in which the double bond is shifted to the exocyclic α-position [18]. Enamines **4** exhibit umpolung reactivity and react with nucleophiles in the presence of Lewis acids (LA) to give α-substituted cyclic oxime ethers **5** (1,2-oxazines or isoxazolines) via S$_N$' substitution of TMSO-group (Scheme 1c). Using this approach, nucleophilic halogenation [19], oxygenation [20–22], azidation [23] of cyclic nitronates were performed (route 1). Although we succeeded in using this methodology in the total synthesis of some pharmaceutical molecules, in the case of nitronates having acid-sensitive groups (e.g., acetals), it proved to be not very efficient (vide infra) [20,24,25].

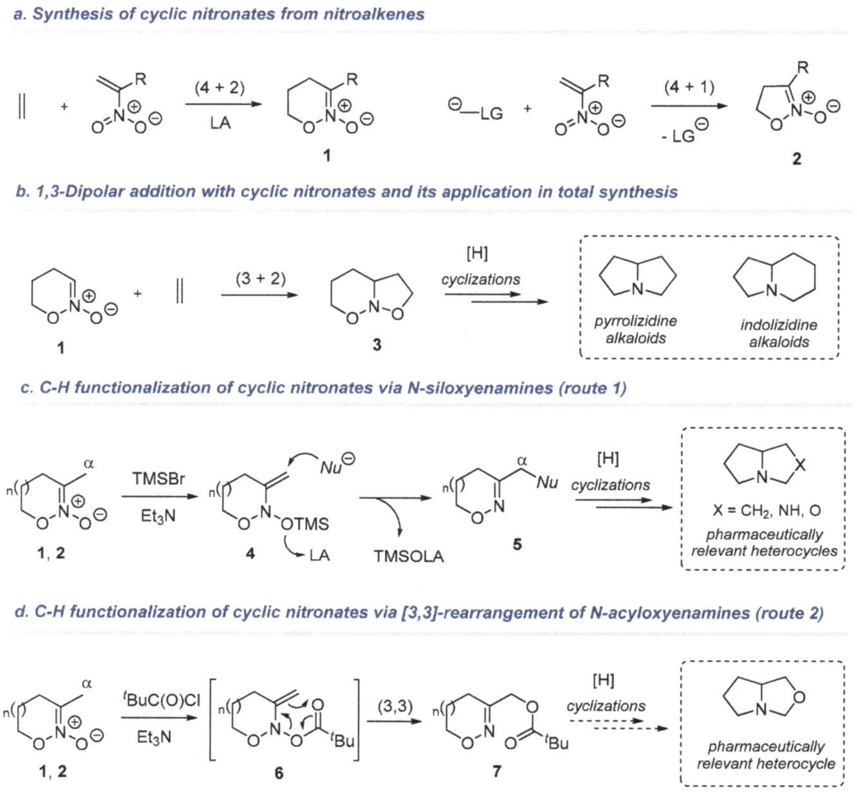

Scheme 1. Approaches towards modification of cyclic nitronates. (**a**) Synthesis of cyclic nitronates from nitroalkenes; (**b**) 1,3-Dipolar addition with cyclic nitronates and its application in total synthesis; (**c**) C-H functionalization of cyclic nitronates via *N*-siloxyenamines (route 1); (**d**) C-H functionalization of cyclic nitronates via [3,3]-rearrangement of *N*-acyloxyenamines (route 2).

Recently, we designed another strategy for the site-selective functionalization of cyclic nitronates **1** and **2** via acylation with acyl halides/Et$_3$N (Scheme 1d) [26]. The initially formed *N*-acyloxyenamines **6** undergo a spontaneous (3,3)-rearrangement to give α-acyloxy-substituted cyclic oxime ethers **7** (route 2). This tandem C–H oxygenation process could be performed under very mild conditions. At present, we are testing the scope and limitation of this method in the total synthesis of some model target molecules to compare its efficacy with our previous C–H functionalization methods. In this study, a potent GlaxoSmithKline's phosphodiesterase 4 (PDE4) inhibitor **CMPO** [27–29], which was previously synthesized using route 1 [24,30], was chosen as a target molecule.

Scheme 2 depicts our previous synthetic route to **CMPO**. The fused pyrrolidine core was prepared by carbamylation of a *trans*-3-aryl-substituted prolinol **15**. The latter was accessed in our strategy by the reductive contraction of the 1,2-oxazine ring in the intermediate **14**, which was prepared by the stereoselective hydride reduction of the 5,6-dihydro-4*H*-1,2-oxazine **13** [24]. This intermediate was synthesized by the reduction of the ONO$_2$-group in the corresponding nitroxy-substituted 1,2-oxazine **12**. Introduction of the nitroxy-group was accomplished by the LA-assisted functionalization of the methyl group in the *N*-oxide **10** using route 1 described above (see Scheme 1c). The required *N*-oxide **10** was assembled in a stereoselective manner by the (4 + 2)-cycloaddition of nitroalkene **8** with the vinyl ether **9** bearing Whitesell's chiral auxiliary group ((+)-*trans*-2-phenyl-1-cyclohexanol ether).

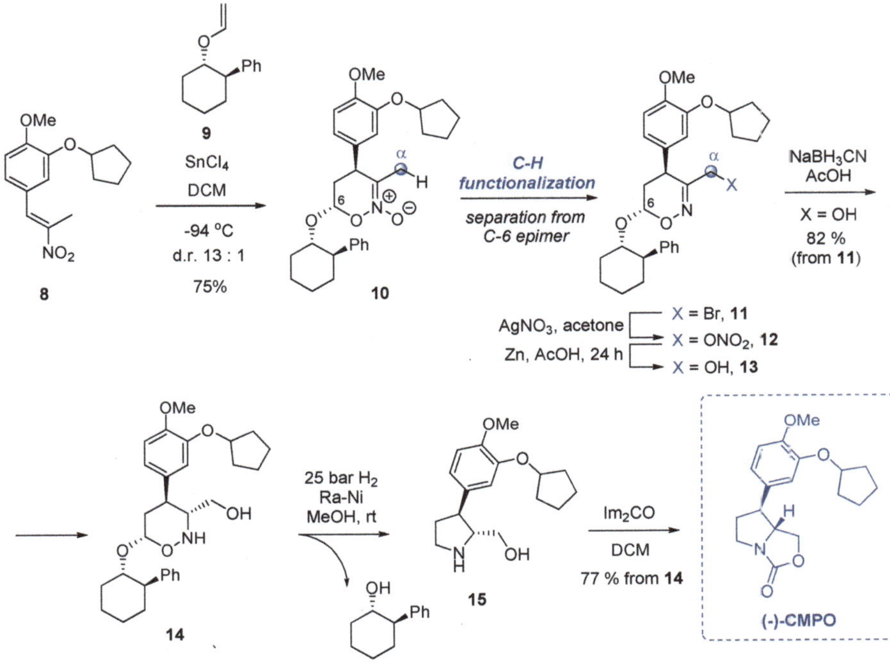

Scheme 2. Previous asymmetric synthesis of (−)-**CMPO**.

Oxygenation of the methyl group in the *N*-oxide **10** proved to be challenging, and several methods were tested (Scheme 3). *N*-Siloxy,*N*-oxyenamine **16** was generated from the *N*-oxide **10** under mild conditions and then subjected to the LA-assisted nucleophilic addition of the bromide anion [24,25]. However, the desired 3-bromomethyl-1,2-oxazine **11**, which served as a precursor to nitrate **12**, was formed in moderate yields (best results are shown in Scheme 3a). Another issue was the epimerization of the sensitive C-6 acetal stereocenter leading to a mixture of 4,6-*trans*/4,6-*cis*-diastereomers **11** and **11'**, which had to be separated by column chromatography (Scheme 3a,b). Unfortunately, the epimer **11'** could not be used in the synthesis of **CMPO** as it produced the undesired 3,4-*cis*-stereoisomer upon the reduction of the C=N bond in 5,6-dihydro-4H-1,2-oxazine ring on the later stages of the synthesis (Scheme 3c) [25].

The reason for the epimerization may lie in the mechanism of the LA-promoted reaction of *N*-oxyenamines with nucleophiles, which involves heterolytic cleavage of the N–O bond (Scheme 3b). Experimental [20,22] and computational data [22] suggest that the S_N' substitution of the TMSO-group may proceed through an epimerizable *N*-vinyl,*N*-oxynitrenium cation **C1**. In our later study, we were able to optimize the epimerization ratio to 6: 1 by using $Cr(NO_3)_3$ as both a mild Lewis acid and the source of the nitrate anion [20]. However, the yield of nitroxy-derivative **12** was still not very high (ca. 40% from nitronate **10**, Scheme 3a). Thus, further optimization of the C–H functionalization stage was reasonable.

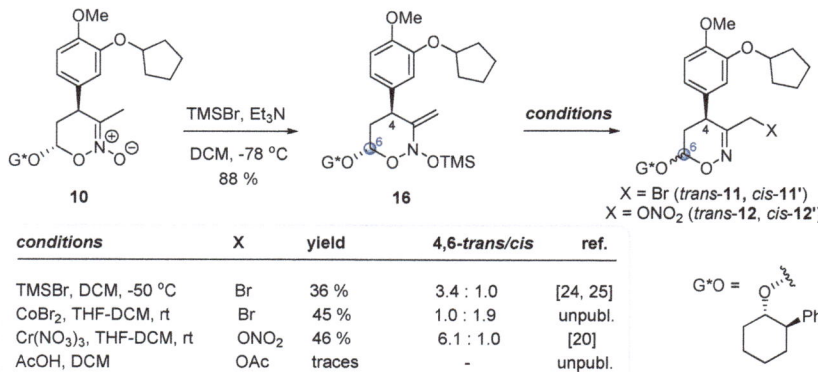

Scheme 3. Problems associated with C-6 epimerization in the LA-assisted functionalization of cyclic nitronate **10**. (**a**) Functionalization of cyclic nitronate **10**; (**b**) Mechanism of epimerization at C-6; (**c**) Hydride reduction of 4,6-*cis*-1,2-oxazines.

2. Results

We speculated that the pericyclic (3,3)-rearrangement of *N*-acyloxyenamine intermediate **I1** generated by the acylation of nitronate **10** may proceed without any epimerization of the C-6 stereogenic center. To test this idea, cyclic nitronate **10** was treated with pivaloyl chloride/Et$_3$N (1.5/2.0 equiv.) under conditions previously optimized for model 1,2-oxazine-*N*-oxides (MeCN, −30 °C, 2 h) [26]. The desired pivalate **17** was formed in a 61% yield together with some amount of unreacted *N*-oxide **10**. After a short optimization of conditions, we found that the use of a bigger access of the PivCl/Et$_3$N system (2.0/2.5 equiv.) and prolonged reaction time (18 h) at lower temperature resulted in an increase in the yield up to 76% (Scheme 4). Gratifyingly, no noticeable epimerization at the C-6 position was observed under these conditions.

We further investigated whether pivalate **17** could be used in the synthesis of **CMPO**. Hydrogenolysis of the pivalate group in **17** (to give alcohol **13**) prior the reduction is challenging since 5,6-dihydro-4*H*-1,2-oxazines are known to undergo fragmentation via a retro-[4+2]-cycloaddition process under the action of bases [31]. Therefore, pivalate **17** was subjected to the hydride reduction with NaBH$_3$CN in acetic acid (Scheme 5). Surprisingly, the reaction produced two separable isomeric products **18** and **18'** in 3: 1 ratio (62% combined yield, 91% based on converted **17**). From the coupling constants in ^{1}H-NMR spectra, it was deduced that both isomers had *trans*-arrangement of the substituents at the C-3 and C-4 atoms, while the configuration of the C-6 stereocenter was different. Thus, the C-6 acetal moiety underwent epimerization in the course of the reduction (see Discussion section). The amount of 4,6-*cis*-isomer **18'** increased with time, demonstrating that the isomerization took place in the reduced product **18** and not in the starting compound **17**. This is also confirmed by the fact that stereisomers **18** and **18'** had same configuration of the newly formed C-3 stereocenter. If the epimerization preceded the reduction, the C-6 epimerized 5,6-dihydro-4H-1,2-oxazine **17'** would also give the reduced product with the 3,4-*cis*-disposition of substituents (see Scheme 5c).

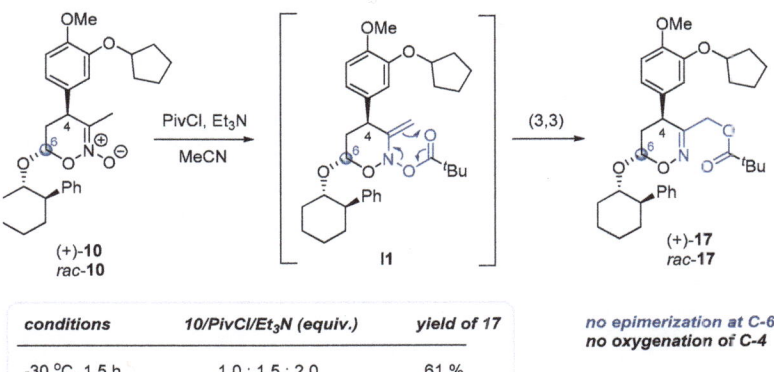

Scheme 4. Tandem acylation/(3,3)-rearrangement of nitronate **10**.

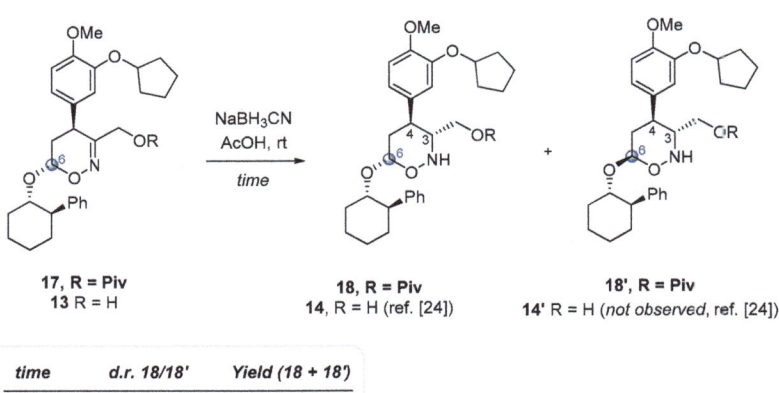

Scheme 5. Cyanoborohydride reduction of 3-hydroxymethyl 1,2-oxazine **13** and its pivalate **17**.

Isomers **18** and **18'** could be almost equally used in the reductive contraction of the 1,2-oxazine ring [32], since both produced the same amino aldehyde intermediate **I3** upon cleavage of the N–O bond followed by fragmentation of the hemiacetal **I2** (Scheme 6). Subsequent intramolecular reductive amination in the intermediate **I3** and protection with the Boc-group afforded the desired prolinol ester **19**. Hence, the separation of epimers **18** and **18'** was not required and a mixture could be converted into the product **19** in a 61% yield. The chiral auxiliary alcohol (*trans*-2-phenylcyclohexanol) was recovered at this stage in 77% yield.

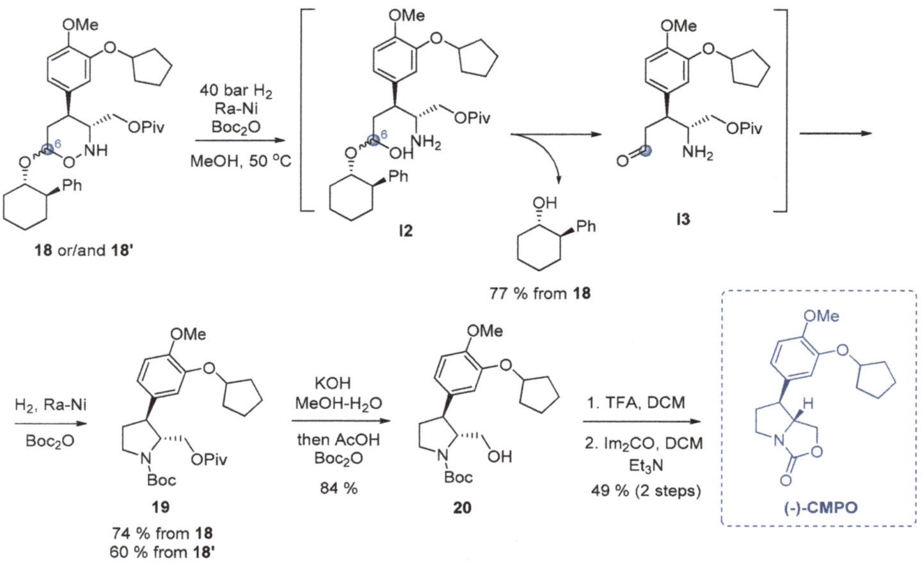

Scheme 6. Conversion of 1,2-oxazines **18** and **18'** into target PDE4 inhibitor (−)-**CMPO**.

On the next step, careful saponification of the pivalate moiety in pyrrolidine **19** with KOH in aqueous methanol gave Boc-prolinol **20** in 84% yield (Scheme 6). It is noteworthy that hydrolysis of the Boc-group was also observed to some extent under these conditions. For this reason, the reaction mixture was treated with Boc$_2$O after neutralization to convert the unprotected prolinol into the N-Boc-derivative **20**, which was isolated by column chromatography.

Finally, deprotection of the N-Boc moiety with TFA and treatment of the resulting prolinol trifluoroacetate with Im$_2$CO/Et$_3$N afforded the desired PDE4 inhibitor **CMPO** (Scheme 6). Thus, the asymmetric synthesis of PDE4 inhibitor **CMPO** was completed in seven steps from a known nitroalkene **8** in 8% overall yield. Chiral HPLC analysis revealed the enantiomeric purity of the product >97% *ee*. The racemic sample of **CMPO** for HPLC analysis was prepared according to the same synthetic sequence starting from the racemic *trans*-2-phenylcyclohexanol.

3. Discussion

Epimerization of the acetal moiety in the course of the hydride reduction of 5,6-dihydro-4-H-1,2-oxazine **17** is of special note. In the previously reported hydride reduction of 1,2-oxazine **13** possessing a free hydroxymethyl group, no epimerization at the C-6 atom was observed (cf. data in Scheme 2; Scheme 5) [24]. We hypothesized that such a difference in the behavior of 3-hydroxymethyl- and 3-acyloxymethyl-substituted 1,2-oxazines **13** and **17** may be attributed to an anchimeric assistance from the carbonyl group, which stabilizes the intermediate cation **C2** [33] by forming a bridged system with an eight-membered ring (cation **C3**). In carbohydrates, a similar anchimeric assistance of the acyloxy group from the distant 1,4-position has been proposed, yet it was not confirmed unambiguously by experimental data [34–37]. In our case, density functional theory (DFT) calculations at the MN15/Def2TZVP level of theory (see Supplementary material for details) revealed that the bridged cation **C3** is much less stable compared to the initial monocyclic cation **C2**. Interestingly, the formation of a third ring between the nitrogen atom and the acyloxy group may lead to a tricyclic cation **C4**, which is predicted to be much more stable than the mono- or bicyclic structures **C2** and **C3** (Scheme 7). The formation of such a stable tricyclic cation as an intermediate or a resting state may account for the observed facile epimerization of pivalate **18**. The higher thermodynamic stability of the 4,6-*cis*-isomer **18′** over the 4,6-*trans*-isomer **18**, as shown by DFT calculations, is likely to be the driving force for the epimerization at the C-6 atom.

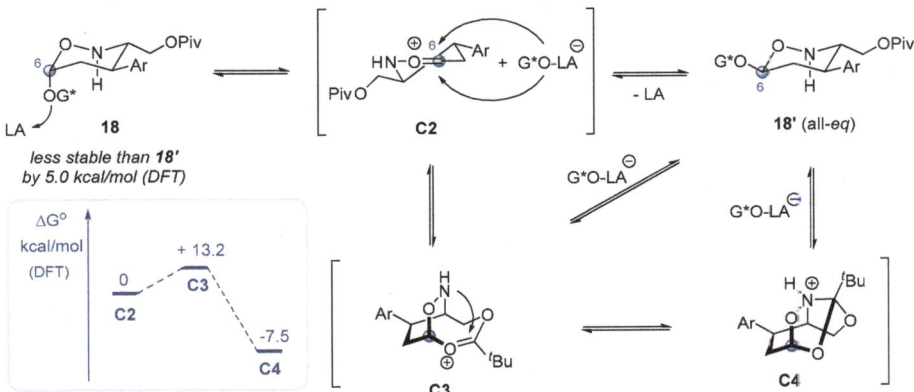

Scheme 7. Plausible mechanism of the epimerization at the C-6 position in 1,2-oxazine **18**. Ar − 4-methoxy-3-cyclopentyloxyphenyl, G*—*trans*-2-phenylcyclohexyl. DFT calculations: MN15/Def2TZVP (AcOH, smd), LA was not included in the calculations.

Another remarkable observation was that 1,2-oxazine **18** (as well as its precursor **17**) did not undergo epimerization in acetic acid (rt, 2 h). Isomerization to the *cis*-isomer **18′** was observed only in the presence of NaBH$_3$CN. Hence, the fragmentation of the acetal moiety is most likely promoted by some Lewis acidic boron species generated from NaBH$_3$CN in acidic medium. Indeed, slow epimerization of 1,2-oxazine **18** was observed upon treatment of **18** with B(OBu)$_3$ or BF$_3$·Et$_2$O in acetic acid.

It is noteworthy that an anchimeric-assisted epimerization in 1,2-oxazine series has not been reported previously. Moreover, to our knowledge, this is the first reported example of a remote neighboring group participation from the 1,4-position in a six-membered ring confirmed by DFT calculations [38–40]. The formation of an eight-membered ring in this case may be driven by an unusual secondary anchimeric interaction involving the nitrogen atom of the 1,2-oxazine ring leading to an unusual tricyclic oxazolo(1,2)oxazinium cation **C4**.

4. Materials and Methods

All reactions were carried out in oven-dried (150 °C) glassware. NMR spectra (Bruker AM 300 spectrometer, Karlsruhe, Germany) were recorded at room temperature (if not stated otherwise) with residual solvents peaks as an internal standard. Peak multiplicities are indicated by s (singlet), d (doublet), t (triplet), dd (doublet of doublets), q (quartet), quint (quintet), ddd (doublet of doublets of doublets), tt (triplet of triplets), tdd (triplets of doublets of doublets), m (multiplet), br (broad). The numeration of atoms used in the assignment of NMR spectra is given in Figure 1.

HRMS were measured on the electrospray ionization (ESI) (Bruker MicroTOF, Karlsruhe, Germany) instrument with a time-of-flight (TOF) detector. Concentrations c in optical rotation angles are given in g/100 mL. [α]$_D$ values are given in 10^{-1} deg cm^2 g^{-1}. Column chromatography was performed using Kieselgel (Merck, Germany) 40–60 μm 60A. Analytical thin-layer chromatography was performed on silica gel plates with QF 254. Visualization was accomplished with UV light and solution of anisaldehyde/H$_2$SO$_4$ in ethanol. Chiral HPLC analysis was performed on a chromatograph with a UV-VIS photodiode array detector (Shimadzu LC-20, Shimadzu, Japan). All reagents were commercial grade and used as received. Et$_3$N, dichloromethane (DCM), and MeCN were distilled over CaH$_2$ prior to the experiments; MeOH, hexane and ethyl acetate were distilled without drying agents.

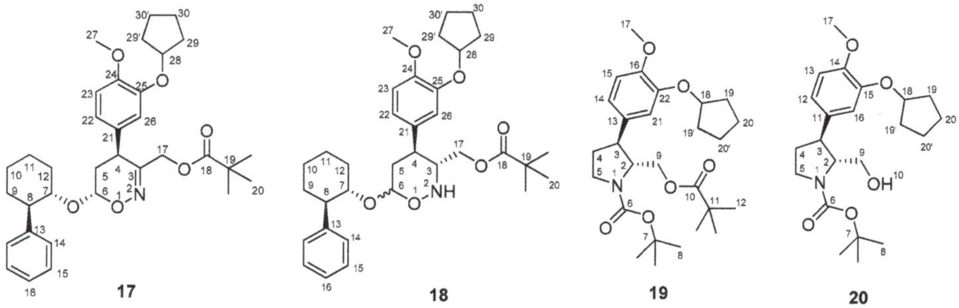

Figure 1. Numeration of atoms in products **17–20**.

(+)-(4S,6S)-4-(3-(Cyclopentyloxy)-4-methoxyphenyl)-3-methyl-6-(((1S,2R)-2-phenylcyclohexyl)oxy)-5, 6-dihydro-4H-1,2-oxazine 2-oxide (+)-10 and the corresponding racemate rac-**10** were synthesized according to the previously reported procedure from nitroalkene **8** and vinyl ether **9** [25]. In most experiments, very similar yields in enantiomeric and racemic series were obtained. In Schemes and Procedures, the best yields among two series are given if not otherwise stated.

Oxygenation of nitronate **10**. Enantipure or racemic nitronate **10** (347 mg, 0.72 mmol) was dissolved in dry acetonitrile (1.7 mL) in a Schlenk tube under argon atmosphere, and then Et$_3$N (251 µL, 1.8 mmol) was added. The solution was cooled to −40 °C and pivaloyl chloride (174 µL, 1.41 mmol) was added. The reaction mixture was stirred at ca. −40 °C for 2 h and then kept in a freezer (ca. −25 °C) overnight. The mixture was diluted with EtOAc (5 mL) and transferred into a separating funnel containing EtOAc (20 mL) and 0.25 M aq. NaHSO$_4$ solution (20 mL). The aqueous layer was extracted with EtOAc (20 mL), the combined organic layers were washed with water (30 mL) and brine (30 mL), dried over anhydrous Na$_2$SO$_4$ and concentrated under reduced pressure. The residue was subjected to a column chromatography on silica gel (Hexane/EtOAc = 10/1) to give 311 mg (76%) of enantiopure or racemic pivalate **17**.

(1S,2R,4S,6S)- and *(1S*,2R*,4S*,6S*)-(4-(3-(cyclopentyloxy)-4-methoxyphenyl)-6-((2-phenylcyclohexyl) oxy)-5,6-dihydro-4H-1,2-oxazin-3-yl)methyl pivalate* (**17**). R$_f$ = 0.44 (hexane/EtOAc = 3/1). ^{1}H-NMR (300 MHz, COSY, HSQC, CDCl$_3$) δ 7.41–7.13 (m, 5H, H14, H15, H16), 6.75 (d, *J* = 8.2 Hz, 1H, H23), 6.57 (dd, *J* = 8.2, 2.1 Hz, 1H, H22), 6.53 (d, *J* = 2.2 Hz, 1H, H26), 5.36 (dd, *J* = 2.7, 2.5 Hz, 1H, H$_{eq}$6), 4.71 (m, 1H, H28), 4.08 (d, *J* = 13.5 Hz, 1H, H17'), 4.02 (ddd, *J* = 10.3, 10.2, 4.0 Hz, 1H, H$_{ax}$7), 3.98 (d, *J* = 13.5 Hz, 1H, H17''), 3.80 (s, 3H, H27), 2.92 (dd, *J* = 11.9, 7.7 Hz, 1H, H$_{ax}$4), 2.61 (ddd, *J* = 11.1, 10.3, 3.6 Hz, 1H, H$_{ax}$8), 2.37 (m, 1H, H12), 2.01 (ddd, *J* = 13.0, 7.7, 2.7 Hz, 1H, H$_{eq}$5), 1.98–1.73 (m, 10H, H$_{ax}$5, H9', H10, H29, H30'), 1.70–1.52 (m, 3H, H9'', H30''), 1.50–1.28 (m, 3H, H11, H$_{eq}$12), 1.10 (s, 9H, H20). ^{13}C-NMR (75 MHz, HSQC, CDCl$_3$) δ 177.4 (C18), 155.3 (C3), 149.3 and 147.9 (C24 and C25), 144.4 (C13), 131.1 (C21), 128.0 and 127.8 (2 C14 and 2 C15), 125.9 (C16), 120.7 (C22), 114.9 (C26), 112.2 (C23), 91.0 (C6), 80.4 (C28), 76.1 (C7), 63.6 (C17), 56.1 (C27), 50.7 (C8), 38.6 (C19), 34.3 (C9), 33.8 (C4), 32.8 and 32.7 (C29 and C29'), 32.5 (C5), 30.6 (C12), 27.1 (3 C20), 26.1 (C11), 24.7 (C10), 24.0 (C30 and C30'). HRMS (ESI): *m/z* calcd. for [C$_{34}$H$_{46}$NO$_6$]$^+$ 564.3320, found 564.3316 [M + H]$^+$.

(+)-*(1S,2R,4S,6S)*-**17**. Colorless oil. [α]$_D$ = +188.6 (c = 0.09, EtOAc, 20 °C). *rac-***17**. Colorless oil.

Hydride reduction of 5,6-dihydro-4H-1,2-oxazine **17**. *Procedure 1:* Enantiopure or racemic pivalate **17** (80 mg, 0.14 mmol) was dissolved in acetic acid (0.8 mL) and sodium cyanoborohydride (120 mg, 1.9 mmol) was added to the solution upon intensive stirring. The reaction mixture was stirred under argon for 30 min at rt, then diluted with EtOAc (3 mL) and transferred into a separating funnel containing EtOAc (20 mL) and a sat. aq. NaHCO$_3$ solution (20 mL). The aqueous layer was extracted with EtOAc (20 mL). The combined organic layers were washed with sat. aq. NaHCO$_3$ solution (20 mL) and brine (40 mL), then dried over Na$_2$SO$_4$ and concentrated under reduced pressure. The residue was subjected to a column chromatography on silica gel (Hexane/EtOAc = 10/1→5/1→3/1) to yield

49 mg (62%) of a mixture of 1,2-oxazines **18** and **18′** (d.r. 3: 1). Also, 26 mg (33%) of unreacted pivalate **17** was isolated from the column chromatography.

Procedure 2: Enantiopure or racemic pivalate **17** (190 mg, 0.34 mmol) was dissolved in acetic acid (1.8 mL) and sodium cyanoborohydride (250 mg, 3.97 mmol) was added to the solution upon intensive stirring. The reaction mixture was stirred under argon at rt for 1.7 h, then diluted with EtOAc (3 mL) and transferred into a separating funnel containing EtOAc (20 mL) and a sat. aq. NaHCO$_3$ solution (20 mL). The aqueous layer was extracted with EtOAc (20 mL). The combined organic layers were washed with sat. aq. NaHCO$_3$ solution (20 mL) and brine (40 mL), then dried over Na$_2$SO$_4$ and concentrated under reduced pressure. The residue was subjected to a column chromatography on silica gel (Hexane/EtOAc = 10/1→5/1→3/1) to yield 48 mg (25%) of the fast moving isomer **18** and 69 mg (36%) of the slow moving isomer **18′**.

((3R,4S,6S)- and (3R,4S*,6S*)-4-(3-(Cyclopentyloxy)-4-methoxyphenyl)-6-(((1S,2R)-2-phenylcyclohexyl) oxy)-1,2-oxazinan-3-yl)methyl pivalate* (**18**). Colorless oil (both enantiopure and racemic). R$_f$ = 0.37 (hexane/EtOAc = 3/1). ^{1}H-NMR (300 MHz, COSY, HSQC, CDCl$_3$) δ 7.55–7.32 (m, 4H, H14, H15), 7.29–7.21 (m, 1H, H16), 6.76 (d, J = 8.0 Hz, 1H, H23), 6.61 (dd, J = 8.0, 2.1 Hz, 1H, H22), 6.59 (br s, 1H, H26), 4.99 (d, J = 3.0 Hz, 1H, H6), 4.79–4.68 (m, 1H, H28), 4.03 (br d, J = 10.7 Hz, 1H, H$_{ax}$2), 3.80 (s, 3H, H27), 3.78 (ddd, J = 10.0, 10.0, 3.9 Hz, 1H, H$_{ax}$7), 3.51 (dd, J = 11.8, 2.8 Hz, 1H, H17′), 3.22 (br ddd, J = 10.7, 9.8, 8.5 Hz, 1H, H$_{ax}$3), 3.08 (dd, J = 11.8, 8.5 Hz, 1H, H17″), 2.68 (ddd, J = 12.3, 10.3, 3.7 Hz, 1H, H$_{ax}$8), 2.32 (ddd, J = 11.9, 11.9, 4.5 Hz, 1H, H$_{ax}$4), 2.21 (br d, J = 10.6 Hz, 1H, H$_{ax}$12), 2.05–1.75 (m, 11H, H5, H9′, H10, H29, H30′), 1.70–1.54 (m, 3H, H9″, H30″), 1.48–1.25 (m, 3H, H11, H$_{eq}$12), 1.12 (s, 9H, H20). ^{13}C-NMR (75 MHz, HSQC, HMBC, DEPT135, CDCl$_3$) δ 178.1 (C18), 149.1 and 147.8 (C24 and C25), 144.5 (C13), 133.9 (C21), 128.8 and 128.0 (2 C14 and 2 C15), 126.9 (C16), 119.5 (C22), 114.2 (C26), 112.3 (C23), 93.5 (C6), 80.5 (C28), 77.7 (C7), 63.3 (C17), 60.1 (C3), 56.1 (C27), 50.7 (C8), 38.6 (C19), 37.2 (C4), 37.0 (C5), 33.3 (C9), 32.7 and 32.8 (C29 and C29′), 31.3 (C12), 27.1 (3 C20), 26.0 (C11), 24.8 (C10), 24.0 (C30 and C30′). HRMS (ESI): *m/z* calcd. for [C$_{34}$H$_{48}$NO$_6$]$^+$ 566.3476, found 566.3474 [M + H]$^+$.

((3R,4S,6R)- and (3R,4S*,6R*)-4-(3-(Cyclopentyloxy)-4-methoxyphenyl)-6-(((1S,2R)-2-phenylcyclohexyl) oxy)-1,2-oxazinan-3-yl)methyl pivalate* (**18′**). Characterized in mixture with **18** (d.r. **18/18′** = 1: 4). Colorless oil (both enantiopure and racemic). R$_f$ = 0.30 (hexane/EtOAc = 3/1). ^{1}H-NMR (300 MHz, COSY, HSQC, CDCl$_3$) δ 7.55–7.32 (m, 5H, H14, H15, H16), 6.79 (d, J = 8.2 Hz, 1H, H23), 6.60 (dd, J = 8.2, 2.2 Hz, 1H, H22), 6.55 (d, J = 2.2 Hz, 1H, H26), 5.43 (d, J = 10.4 Hz, 1H, H2), 5.28 (d, J = 2.9 Hz, 1H, H6), 4.73 (m, 1H, H28), 3.93 (ddd, J = 10.3, 10.2, 4.0 Hz, 1H, H7), 3.81 (s, 3H, H27), 3.59 (d, J = 12.9 Hz, 1H, H17), 3.48 (ddd, J = 11.2, 10.4, 5.7 Hz, 1H, H$_{ax}$3), 3.38 (dd, J = 12.9, 5.7 Hz, 1H, H17), 2.71 (ddd, J = 10.6, 10.3, 3.3 Hz, 1H, H$_{ax}$8), 2.63 (ddd, J = 12.5, 11.2, 4.0 Hz, 1H, H$_{ax}$4), 2.24 (br d, J = 11.9 Hz, 1H, H$_{ax}$12), 2.06 (td, J = 13.3, 12.5, 2.9 Hz, 1H, H$_{ax}$5), 1.99–1.76 (m, 10H, H$_{eq}$5, H9′, H10, H29, H30′), 1.72–1.53 (m, 3H, H9″, H30″), 1.53–1.25 (m, 3H, H11, H$_{eq}$12), 1.10 (s, 9H, H20). ^{13}C-NMR (75 MHz, HSQC, HMBC, DEPT135, CDCl$_3$) δ 177.4 (C18), 149.9 and 148.2 (C24 and C25), 143.7 (C13), 130.6 (C21), 129.6 (C16), 127.7 and 127.5 (2 C14 and 2 C15), 120.0 (C22), 113.7 (C26), 112.4 (C23), 95.2 (C6), 80.7 (C28), 78.5 (C7), 65.8 (C3), 61.3 (C17), 56.0 (C27), 50.5 (C8), 38.5 (C19), 36.7 (C4), 36.4 (C5), 33.1 (C9), 32.8 and 32.7 (C29 and C29′), 30.7 (C12), 27.0 (3 C20), 25.8 (C11), 24.4 (C10), 24.0 (C30 and C30′). HRMS (ESI): *m/z* calcd. for [C$_{34}$H$_{48}$NO$_6$]$^+$ 566.3476, found 566.3476 [M + H]$^+$.

Hydrogenation of 1,2-oxazines **18** and **18′**. A glass vial was charged with a solution of enantiopure 1,2-oxazine **18** (48 mg, 0.086 mmol) and Boc$_2$O (58 mg, 0.264 mmol) in methanol (0.5 mL). A suspension of Raney nickel (ca. 100 mg, prepared from 50% slurry in water) in methanol (ca. 0.5 mL) was added, and the vial was placed in a steel autoclave, which was then flushed and filled with hydrogen to a pressure of ca. 40 bar and heated to 50 °C. The hydrogenation was conducted for 2 h with intensive stirring. Then, the autoclave was cooled to rt, slowly depressurized, and the catalyst was removed using a magnet and washed with methanol. The solution was concentrated to dryness under reduced pressure. The residue was subjected to a column chromatography on silica gel (hexane/EtOAc = 20/1→10/1) to yield 31 mg (74%) of *N*-Boc pyrrolidine **19**. The column was then washed with hexane/EtOAc = 5/1 to recover (+)-*trans*-2-phenylcyclohexanol (12 mg, 77%).

Application of the same procedure for the reduction of enantiopure 1,2-oxazine **18′** (69 mg, 0.13 mmol) afforded 36 mg (60%) of N-Boc pyrrolidine **19** and 15 mg (66%) of (+)-*trans*-2-phenylcyclohexanol.

Application of the same procedure for the reduction of a mixture of racemic 1,2-oxazines *rac*-**18** and *rac*-**18′** (90 mg, 0.159 mmol, d.r. 1: 1.4) afforded 46 mg (61%) of racemic N-Boc pyrrolidine *rac*-**19**.

Tert-butyl (2R,3S)- and (2R*,3S*)-3-(3-(cyclopentyloxy)-4-methoxyphenyl)-2-((pivaloyloxy)methyl) pyrrolidine-1-carboxylate (**19**). R_f = 0.30 (hexane/EtOAc = 3/1). ^{1}H-NMR (300 MHz, 320K, COSY, HSQC, CDCl$_3$) δ 6.82 (d, *J* = 8.7 Hz, 1H, H15), 6.75–6.70 (m, 2H, H14 and H21), 4.76 (m, 1H, H18), 4.41–4.26 (m, 1H, H9′), 4.20 (br d, *J* = 9.9 Hz, 1H, H9″), 4.11–3.88 (br m, 1H, H2), 3.84 (s, 3H, H17), 3.77–3.62 (br m, 1H, H5′), 3.46–3.31 (m, 1H, H5″), 3.29–3.10 (br m, 1H, H3), 2.35–2.18 (m, 1H, H4′), 2.01–1.76 (m, 7H, H4″, H19, H20′), 1.72–1.54 (m, 2H, H20″), 1.49 (s, 9H, H8), 1.21 (s, 9H, H12). ^{13}C-NMR (75 MHz, HSQC, DEPT135, CDCl$_3$) δ 178.2 (C10), 154.2 (C6), 148.9 and 147.9 (C16 and C22), 135.1 (C13), 119.2 (C14), 114.0 (C21), 112.3 (C15), 80.5 (C18), 80.0 (C7), 63.5 (br, C9), 63.1 (br, C2), 56.2 (C17), 47.2 (C3), 46.3 (C5), 38.8 (C11), 32.8 (C19 and C19′), 31.8 (C4), 28.5 (3 C8), 27.2 (3 C12), 24.0 (C20 and C20′) (signals are broadened due to the presence of N-Boc rotamers). HRMS (ESI): *m/z* calcd. for $[C_{27}H_{42}NO_6]^+$ 476.3007, found 476.3006 $[M + H]^+$.

(−)-(2R,3S)-**19**. Colorless oil, $[α]_D$ = −19.0 (*c* = 1, EtOAc, 24 °C). *rac*-**19**. Colorless oil.

Saponification of pivalate **19**. Enantiopure or racemic pivalate **19** (67 mg, 0.14 mmol) was dissolved in MeOH (2.8 mL) and a solution of KOH (237 mg, 4.2 mmol) in H$_2$O (1.4 mL) was added. The mixture was stirred at room temperature for 24 h. Then, acetic acid (0.4 mL) was added and the reaction mixture was stirred for 5 min. The resulting solution was concentrated in vacuum. To the residue, MeOH (2 mL) and Boc$_2$O (0.065 g, 0.28 mmol) were added and the resulting solution was stirred for 1 h. Then, volatiles were removed in vacuum and the residue was subjected to a column chromatography on silica gel (hexane/EtOAc = 5/1→3/1→1/1) to yield 46 mg (84%) of prolinol **20**.

Tert-butyl (2R,3S)- and (2R*,3S*)-3-(3-(cyclopentyloxy)-4-methoxyphenyl)-2-(hydroxymethyl)pyrrolidine-1-carboxylate (**20**). R_f = 0.21 (hexane/EtOAc = 1/1). ^{1}H-NMR (300 MHz, COSY, HSQC, CDCl$_3$) δ 6.81 (d, *J* = 7.6 Hz, 1H, H13), 6.76 (d, *J* = 7.6 Hz, 1H, H12), 6.74 (s, 1H, H16), 5.13–4.88 (br, 1H, H10), 4.77 (m, 1H, H18), 3.97–3.85 (br m, 1H, H2), 3.83 (s, 3H, H17), 3.80–3.66 (br m, 2H, H9′, H5′), 3.63 (dd, *J* = 11.5, 6.8 Hz, 1H, H9″), 3.44–3.27 (br m, 1H, H5″), 2.89–2.79 (br m, 1H, H3), 2.22–2.07 (br m, 1H, H4′), 2.02–1.75 (m, 7H, H20′, H19, H4″), 1.67–1.57 (m, 2H, H20″), 1.51 (s, 9H, H8). ^{13}C-NMR (75 MHz, HSQC, DEPT135, CDCl$_3$) δ 156.8 (C6), 149.2 and 147.9 (C14 and C15), 133.3 (C11), 119.7 (C12), 114.5 (C16), 112.3 (C13), 80.5 (C18), 80.4 (C7), 67.1 (C2), 66.1 (C9), 56.2 (C17), 47.4 (C3), 47.0 (C5), 32.9 and 32.8 (C-4, C19 and C19′), 28.5 (3 C8), 24.0 (C20 and C20′). HRMS (ESI): *m/z* calcd. for $[C_{22}H_{33}NO_5Na]^+$ 414.2251, found 414.2247 $[M + Na]^+$.

(−)-(2R,3S)-**20**. Colorless oil, $[α]_D$ = −9.3 (*c* = 1, EtOAc, 24 °C). *rac*-**20**. Colorless oil.

Synthesis of (−)- and *rac*-**CMPO**. To a stirred solution of enantiopure or racemic prolinol **20** (45 mg, 0.115 mmol) in DCM (0.9 mL) was added CF$_3$COOH (0.18 mL, 2.4 mmol) at 0–5 °C. The cooling bath was removed, and the solution was stirred for 1 h. Then, volatiles were removed under reduced pressure and the residue was dried until constant weight. The resulting trifluoroacetate was dissolved in DCM (0.85 mL), and Et$_3$N (0.08 mL, 0.58 mmol) and 1,1′-carbonyldiimidazole (47 mg, 0.29 mmol) were added at rt. The solution was stirred for 18 h at rt, and then concentrated under reduced pressure. The product was isolated by column chromatography on silica gel (hexane/EtOAc = 3/1) followed by recrystallization from hexane\diethyl ether (ca. 1: 1). Yield: 18 mg (49%). ^{1}H NMR spectra were in agreement with previously published data [24].

(−)-(7S,7aR)-**CMPO**. White solid. Mp = 134–137 °C (lit.[24] 137–139 °C). HPLC analysis: *ee* > 97% (RT 9.8 min; column CHIRALPAK IA-3 (15 cm); solvent Hexane/*i*-PrOH = 90:10; temperature 40 °C; flow rate 1 mL/min). $[α]_D$ = −63.0 (*c* = 0.5, EtOAc, 25 °C). lit.[24] $[α]_D$ = −69.1 (*c* = 0.83, MeOH, 26 °C).

rac-**CMPO**. White solid. Mp = 103–104 °C (lit.[30] 99–101 °C). HPLC analysis: (+)-(7R,7aS)-**CMPO** (RT 8.8 min) and (−)-(7S,7aR)-**CMPO** (RT 9.8 min); column CHIRALPAK IA-3 (15 cm); solvent Hexane/*i*-PrOH = 90:10; temperature 40 °C; flow rate 1 mL/min.

5. Conclusions

In conclusion, we were able to solve the problem of site-selective C–H oxygenation of the cyclic nitronate intermediate in the asymmetric synthesis of a potent PDE4 inhibitor **CMPO** by using tandem acylation/(3,3)-sigmatropic rearrangement. In comparison with the previous synthesis, this method afforded the required 3-oxymethyl-substituted 1,2-oxazine intermediate in a much higher yield (76% vs. 27%). This key intermediate could be readily converted into the target (−)-**CMPO** by the reductive contraction of the 1,2-oxazine ring followed by deprotection and carbamylation with Im_2CO. A rapid epimerization of the C-6 acetal moiety was observed upon the reduction of the 5,6-dihydro-4H-1,2-oxazine ring with $NaBH_3CN$ in acetic acid. DFT calculations suggest that the epimerization is favored by an unprecedented double anchimeric assistance from a remote acyloxy group and the nitrogen atom of the 1,2-oxazine ring.

Supplementary Materials: The following are available online: NMR spectra for compounds **10, 17, 18, 18′, 19, 20** and **CMPO**, chiral phase HPLC chromatograms for *rac*-**CMPO** and (−)-**CMPO**, Cartesian coordinates, absolute energies for all optimized geometries.

Author Contributions: E.V.P. carried out synthesis, separation, and purification of compounds; I.S.G. performed DFT calculations; S.L.I. collaborated in the discussion and interpretation of the results, manuscript editing; A.Y.S. supervised the whole research, including conceptualization, methodology, data curation and original draft preparation. All authors have read and agreed to the published version of the manuscript.

Funding: This research was funded by the Russian Science Foundation (grant 17-13-01411).

Conflicts of Interest: The authors declare no conflict of interest. The funders had no role in the design of the study; in the collection, analyses, or interpretation of data; in the writing of the manuscript, or in the decision to publish the results.

References

1. Denmark, S.E.; Thorarensen, A. Tandem [4 + 2]/[3 + 2] Cycloadditions of Nitroalkenes. *Chem. Rev.* **1996**, *96*, 137–166. [CrossRef] [PubMed]
2. Tabolin, A.A.; Sukhorukov, A.Y.; Ioffe, S.L.; Dilman, A.D. Recent Advances in the Synthesis and Chemistry of Nitronates. *Synthesis* **2017**, *49*, 3255–3268. [CrossRef]
3. Mukaijo, Y.; Yokoyama, S.; Nishiwaki, N. Comparison of Substituting Ability of Nitronate versus Enolate for Direct Substitution of a Nitro Group. *Molecules* **2020**, *25*, 2048. [CrossRef] [PubMed]
4. de Carvalho, L.L.; Burrow, R.A.; Pereira, V.L.P. Diastereoselective synthesis of nitroso acetals from (S,E)-γ-aminated nitroalkenes via multicomponent [4 + 2]/[3 + 2] cycloadditions promoted by LiCl or $LiClO_4$. *Beilstein J. Org. Chem.* **2013**, *9*, 838–845. [CrossRef] [PubMed]
5. Kano, T.; Yamamoto, A.; Song, S.; Maruoka, K. Catalytic asymmetric syntheses of isoxazoline-N-oxides under phase-transfer conditions. *Chem. Commun.* **2011**, *47*, 4358–4360. [CrossRef] [PubMed]
6. Koc, E.; Kwon, O. Total syntheses of heliotridane and pseudoheliotridane through nitrodiene–acrylate 6π-electrocyclization/[3+2] cycloaddition. *Tetrahedron* **2017**, *73*, 4195–4200. [CrossRef]
7. Creech, G.S.; Kwon, O. Tandem 6π-Electrocyclization and Cycloaddition of Nitrosodienes to Yield Multicyclic Nitroso Acetals. *J. Am. Chem. Soc.* **2010**, *132*, 8876–8877. [CrossRef]
8. Zhu, C.-Y.; Deng, X.-M.; Sun, X.-L.; Zheng, J.-C.; Tang, Y. Highly enantioselective synthesis of isoxazoline N-oxides. *Chem. Commun.* **2008**, *6*, 738–740. [CrossRef]
9. Streitferdt, V.; Haindl, M.H.; Hioe, J.; Morana, F.; Renzi, P.; von Rekowski, F.; Zimmermann, A.; Nardi, M.; Zeitler, K.; Gschwind, R.M. Unprecedented Mechanism of an Organocatalytic Route to Conjugated Enynes with a Junction to Cyclic Nitronates. *Eur. J. Org. Chem.* **2019**, *2019*, 328–337. [CrossRef]
10. Jiang, H.; Elsner, P.; Jensen, K.L.; Falcicchio, A.; Marcos, V.; Jørgensen, K.A. Achieving Molecular Complexity by Organocatalytic One-Pot Strategies—A Fast Entry for Synthesis of Sphingoids, Amino Sugars, and Polyhydroxylated α-Amino Acids. *Angew. Chem. Int. Ed.* **2009**, *48*, 6844–6848. [CrossRef]
11. Baiazitov, R.Y.; Denmark, S.E. Tandem [4 + 2]/[3 + 2] Cycloadditions. In *Methods and Applications of Cycloaddition Reactions in Organic Syntheses*; Nishiwaki, N., Ed.; John Wiley & Sons: Hoboken, NJ, USA, 2014; pp. 471–550.

12. Denmark, S.E.; Martinborough, E.A. Enantioselective Total Syntheses of (+)-Castanospermine, (+)-6-Epicastanospermine, (+)-Australine, and (+)-3-Epiaustraline. *J. Am. Chem. Soc.* **1999**, *121*, 3046–3056. [CrossRef]
13. Denmark, S.E.; Thorarensen, A.; Middleton, D.S. Tandem [4 + 2]/[3 + 2] Cycloadditions of Nitroalkenes. 9. Synthesis of (−)-Rosmarinecine. *J. Am. Chem. Soc.* **1996**, *118*, 8266–8277. [CrossRef]
14. Denmark, S.E.; Baiazitov, R.Y.; Nguyen, S.T. Tandem double intramolecular [4 + 2]/[3 + 2] cycloadditions of nitroalkenes: Construction of the pentacyclic core structure of daphnilactone B. *Tetrahedron* **2009**, *65*, 6535–6548. [CrossRef] [PubMed]
15. Denmark, S.E.; Montgomery, J.I.; Kramps, L.A. Synthesis, X-ray Crystallography, and Computational Analysis of 1-Azafenestranes. *J. Am. Chem. Soc.* **2006**, *128*, 11620–11630. [CrossRef]
16. Denmark, S.E.; Montgomery, J.I. Synthesis of cis,cis,cis,cis-[5.5.5.4]-1-Azafenestrane with Discovery of an Unexpected Dyotropic Rearrangement. *Angew. Chem. Int. Ed.* **2005**, *44*, 3732–3736. [CrossRef]
17. For a review see: Sukhorukov, A.Y. C-H Reactivity of the α-Position in Nitrones and Nitronates. *Adv. Synth. Catal.* **2020**, *362*, 724–754. [CrossRef]
18. Tishkov, A.A.; Lesiv, A.V.; Khomutova, Y.A.; Strelenko, Y.A.; Nesterov, I.D.; Antipin, M.Y.; Ioffe, S.L.; Denmark, S.E. 2-Silyloxy-1,2-oxazines, a New Type of Acetals of Conjugated Nitroso Alkenes. *J. Org. Chem.* **2003**, *68*, 9477–9480. [CrossRef]
19. Sukhorukov, A.Y.; Kapatsyna, M.A.; Yi, T.L.T.; Park, H.R.; Naumovich, Y.A.; Zhmurov, P.A.; Khomutova, Y.A.; Ioffe, S.L.; Tartakovsky, V.A. A General Metal-Assisted Synthesis of α-Halo Oxime Ethers from Nitronates and Nitro Compounds. *Eur. J. Org. Chem.* **2014**, *2014*, 8148–8159. [CrossRef]
20. Naumovich, Y.A.; Buckland, V.E.; Sen'ko, D.A.; Nelyubina, Y.V.; Khoroshutina, Y.A.; Sukhorukov, A.Y.; Ioffe, S.L. Metal-assisted addition of a nitrate anion to bis(oxy)enamines. A general approach to the synthesis of α-nitroxy-oxime derivatives from nitronates. *Org. Biomol. Chem.* **2016**, *14*, 3963–3974. [CrossRef]
21. Tabolin, A.A.; Lesiv, A.V.; Khomutova, Y.A.; Nelyubina, Y.V.; Ioffe, S.L. Rearrangement of 3-alkylidene-2-siloxy-tetrahydro-1,2-oxazines (ASENA). A new approach toward the synthesis of 3-α-hydroxyalkyl-5,6-dihydro-4H-1,2-oxazines. *Tetrahedron* **2009**, *65*, 4578–4592. [CrossRef]
22. Naumovich, Y.A.; Golovanov, I.S.; Sukhorukov, A.Y.; Ioffe, S.L. Addition of HO-Acids to N,N-Bis(oxy)enamines: Mechanism, Scope and Application to the Synthesis of Pharmaceuticals. *Eur. J. Org. Chem.* **2017**, *2017*, 6209–6227. [CrossRef]
23. Zhmurov, P.A.; Khoroshutina, Y.A.; Novikov, R.A.; Golovanov, I.S.; Sukhorukov, A.Y.; Ioffe, S.L. Divergent Reactivity of In Situ Generated Metal Azides: Reaction with N,N-Bis(oxy)enamines as a Case Study. *Chem. Eur. J.* **2017**, *23*, 4570–4578. [CrossRef] [PubMed]
24. Zhmurov, P.A.; Sukhorukov, A.Y.; Chupakhin, V.I.; Khomutova, Y.V.; Ioffe, S.L.; Tartakovsky, V.A. Synthesis of PDE IV inhibitors. First asymmetric synthesis of two of GlaxoSmithKline's highly potent Rolipram analogues. *Org. Biomol. Chem.* **2013**, *11*, 8082–8091. [CrossRef] [PubMed]
25. Sukhorukov, A.Y.; Boyko, Y.D.; Ioffe, S.L.; Khomutova, Y.A.; Nelyubina, Y.V.; Tartakovsky, V.A. Synthesis of PDE IVb Inhibitors. 1. Asymmetric Synthesis and Stereochemical Assignment of (+)- and (−)-7-[3-(Cyclopentyloxy)-4-methoxyphenyl]hexahydro-3H-pyrrolizin-3-one. *J. Org. Chem.* **2011**, *76*, 7893–7900. [CrossRef] [PubMed]
26. Kokuev, A.O.; Antonova, Y.A.; Dorokhov, V.S.; Golovanov, I.S.; Nelyubina, Y.V.; Tabolin, A.A.; Sukhorukov, A.Y.; Ioffe, S.L. Acylation of Nitronates: [3,3]-Sigmatropic Rearrangement of in Situ Generated N-Acyloxy,N-oxyenamines. *J. Org. Chem.* **2018**, *83*, 11057–11066. [CrossRef]
27. Brackeen, M.F.; Stafford, J.A.; Cowan, D.J.; Brown, P.J.; Domanico, P.L.; Feldman, P.L.; Rose, D.; Strickland, A.B.; Veal, J.M.; Verghese, M. Design and Synthesis of Conformationally Constrained Analogs of 4-(3-Butoxy-4-methoxybenzyl)imidazolidin-2-one (Ro 20-1724) as Potent Inhibitors of cAMP-Specific Phosphodiesterase. *J. Med. Chem.* **1995**, *38*, 4848–4854. [CrossRef]
28. Jackson, E.K.; Carcillo, J.A. Treatment of Sepsis-Induced Acute Renal Failure. U.S. Patent 5849774, 15 December 1998.
29. Mulhall, A.M.; Droege, C.A.; Ernst, N.E.; Panos, R.J.; Zafar, M.A. Phosphodiesterase 4 inhibitors for the treatment of chronic obstructive pulmonary disease: A review of current and developing drugs. *Exp. Opin. Investig. Drugs* **2015**, *24*, 1597–1611. [CrossRef]

30. Zhmurov, P.A.; Tabolin, A.A.; Sukhorukov, A.Y.; Lesiv, A.V.; Klenov, M.S.; Khomutova, Y.A.; Ioffe, S.L.; Tartakovsky, V.A. Synthesis of phosphodiesterase IVb inhibitors 2. Stereoselective synthesis of hexahydro-3H-pyrrolo[1,2-c]imidazol-3-one and tetrahydro-1H-pyrrolo[1,2-c][1,3]oxazol-3-one derivatives. *Russ. Chem. Bull.* **2011**, *60*, 2390–2395. [CrossRef]
31. Yasuo, T.; Norihiko, Y.; Hiroko, M.; Toshihide, K.; Masahiro, A.; Yutaka, M. Synthesis and a Novel Fragmentation of 6-Alkoxy-5,6-dihydro-4H-1,2-oxazine 2-Oxide. *Bull. Chem. Soc. Jpn.* **1988**, *61*, 461–465.
32. Sukhorukov, A.Y.; Ioffe, S.L. Chemistry of Six-Membered Cyclic Oxime Ethers. Application in the Synthesis of Bioactive Compounds. *Chem. Rev.* **2011**, *111*, 5004–5041. [CrossRef]
33. Zimmer, R.; Arnold, T.; Homann, K.; Reissig, H.-U. An Efficient and Simple Synthesis of 3,4,5,6-Tetrahydro-2H-1,2-oxazines by Sodium Cyanoborohydride Reduction of 5,6-Dihydro-4H-1,2-oxazines. *Synthesis* **1994**, *1994*, 1050–1056. [CrossRef]
34. Demchenko, A.V.; Rousson, E.; Boons, G.-J. Stereoselective 1,2-cis-galactosylation assisted by remote neighboring group participation and solvent effects. *Tetrahedron Lett.* **1999**, *40*, 6523–6526. [CrossRef]
35. Ma, Y.; Lian, G.; Li, Y.; Yu, B. Identification of 3,6-di-O-acetyl-1,2,4-O-orthoacetyl-α-d-glucopyranose as a direct evidence for the 4-O-acyl group participation in glycosylation. *Chem. Commun.* **2011**, *47*, 7515–7517. [CrossRef] [PubMed]
36. Beaver, M.G.; Billings, S.B.; Woerpel, K.A. C-Glycosylation Reactions of Sulfur-Substituted Glycosyl Donors: Evidence against the Role of Neighboring-Group Participation. *J. Am. Chem. Soc.* **2008**, *130*, 2082–2086. [CrossRef] [PubMed]
37. Stalford, S.A.; Kilner, C.A.; Leach, A.G.; Turnbull, W.B. Neighbouring group participation vs. addition to oxacarbenium ions: Studies on the synthesis of mycobacterial oligosaccharides. *Org. Biomol. Chem.* **2009**, *7*, 4842–4852. [CrossRef] [PubMed]
38. Yang, B.; Yang, W.; Ramadan, S.; Huang, X. Pre-Activation-Based Stereoselective Glycosylations. *Eur. J. Org. Chem.* **2018**, *2018*, 1075–1096. [CrossRef]
39. Crich, D.; Hu, T.; Cai, F. Does Neighboring Group Participation by Non-Vicinal Esters Play a Role in Glycosylation Reactions? Effective Probes for the Detection of Bridging Intermediates. *J. Org. Chem.* **2008**, *73*, 8942–8953. [CrossRef]
40. Komarova, B.S.; Tsvetkov, Y.E.; Nifantiev, N.E. Design of α-Selective Glycopyranosyl Donors Relying on Remote Anchimeric Assistance. *Chem. Rec.* **2016**, *16*, 488–506. [CrossRef]

Sample Availability: Samples of racemic and enantiopure PDE4 inhibitor **CMPO** are available from the authors.

© 2020 by the authors. Licensee MDPI, Basel, Switzerland. This article is an open access article distributed under the terms and conditions of the Creative Commons Attribution (CC BY) license (http://creativecommons.org/licenses/by/4.0/).

Communication

Synthesis and Properties of NitroHPHAC: The First Example of Substitution Reaction on HPHAC

Yoshiki Sasaki, Masayoshi Takase *[ID], Shigeki Mori[ID] and Hidemitsu Uno *[ID]

Graduate School of Science and Engineering, Ehime University, Matsuyama 790-8577, Japan; f865001a@mails.cc.ehime-u.ac.jp.com (Y.S.); mori.shigeki.mu@ehime-u.ac.jp (S.M.)
* Correspondence: takase.masayoshi.ry@ehime-u.ac.jp (M.T.); uno@ehime-u.ac.jp (H.U.); Tel.: +81-89-927-9612 (M.T.); +81-89-927-9610 (H.U.)

Academic Editor: Nagatoshi Nishiwaki
Received: 4 April 2020; Accepted: 24 May 2020; Published: 27 May 2020

Abstract: Hexapyrrolohexaazacoronene (HPHAC) is one of the N-containing polycyclic aromatic hydrocarbons in which six pyrroles are fused circularly around a benzene. Despite the recent development of HPHAC analogues, there is no report on direct introduction of functional groups into the HPHAC skeleton. This work reports the first example of nitration reaction of decaethylHPHAC. The structures of nitrodecaethylHPHAC including neutral and two oxidized species (radical cation and dication), intramolecular charge transfer (ICT) character, and global aromaticity of the dication are discussed.

Keywords: hexapyrrolohexaazacoronene; nitration; $S_N Ar$ substitution; ICT character; aromaticity

1. Introduction

Introduction of functional groups into π-conjugated systems is a straightforward way to tune their chemical and physical properties. A nitro group is one of the most useful groups for such purpose. Aromatic nitro compounds have been studied and utilized for medicines, pesticides, dyes, pigments, plastics raw materials, and so on. The functional group strongly stabilizes electron-rich polycyclic aromatic hydrocarbons (PAHs) and can be transformed to many other functional groups by simple reactions [1]. When the nitro group is introduced to large π-conjugated compounds, the LUMO levels would be sufficiently lowered so that nucleophiles could directly attack the π-conjugated systems. The nitro group can be easily transformed into an amino group, which is recognized as one of the most powerful electron-donating groups, as well. Therefore, nitration reaction is one of the first choices for further derivatization of aromatic compounds. Recently, large π-conjugated materials with a nitro group and their derivatives have attracted continuing attention from many research areas [2–14].

Hexapyrrolohexaazacoronene (HPHAC) is a nitrogen-embedded PAH, consisting of circularly connected pyrroles around a benzene core and is easily prepared from hexapyrrolylbenzene by the Scholl oxidation using $FeCl_3$ [15], 2,3-dichloro-5,6-dicyano-1,4-benzoquinone (DDQ) in the presence of trifluoromethanesulfonic acid [16], or N-bromosuccinimide (NBS) [17]. The HPHAC π-system consists of local π-systems of six pyrroles and one benzene and has a non-aromatic character in the whole molecule. Due to the pyrrole moieties, on the other hand, multiple oxidation levels of HPHAC are reversibly generated, and its dication shows global aromaticity owing to the cyclic conjugation of a 22π-electron system. So far, HPHAC-hexabenzocoronene-hybridized [18], ethylene-bridged [19], periphery-expanded [20,21], core-expanded [16], azulene-fused [22], β,β-thieno-fused [23], chiral [24], antiaromatic [25], and σ-dimerized [26] types of its analogues have been reported. However, a peripheral substitution reaction of HPHAC has never been reported yet. Functionalization of HPHACs has all been dependent on the composed pyrroles employed. Herein, we report the first example of nitration on decaethylHPHAC (DEHPHAC **1a**), leading to nitroDEHPHAC **2a** (Scheme 1). The fundamental properties of nitroDEHPHAC **2a** were investigated by NMR and absorption spectroscopy, single crystal X-ray structure analysis, and theoretical calculations.

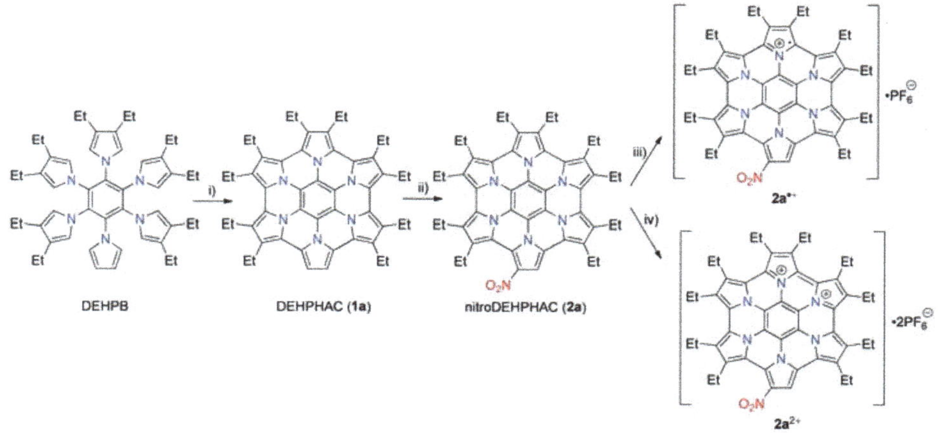

Scheme 1. Synthesis of nitroDEHPHAC **2a** and its oxidized salts **2a**$^{\bullet+}$[PF$_6^-$] and **2a**$^{2+}$[PF$_6^-$]$_2$. Reagents, conditions, and yield: (i) FeCl$_3$, CH$_3$NO$_2$/CH$_2$Cl$_2$, reflux; 67%; (ii) AgNO$_2$ (9.7 equiv.), CH$_2$Cl$_2$, rt; 51%; (iii) AgPF$_6$ (1.0 equiv.), CH$_2$Cl$_2$, rt; quant.; (iv) AgPF$_6$ (2.0 equiv.), CH$_2$Cl$_2$, rt; quant.

2. Results and Discussion

DEHPHAC **1a** was synthesized from *N*-(2,3,4,5,6-pentafluorophenyl)-1*H*-pyrrole via decaethylated hexapyrrolylbenzene DEHPB by S$_N$Ar and successive Scholl reactions in accordance with the previously reported procedure [15,27]. The nitration of **1a** was then carried out with silver nitrite. In contrast to the cases of corrole and porphyrin [28,29], DEHPHAC **1a** did not dimerize by this oxidant but was transformed into nitroDEHPHAC **2a** in a 51% yield (Scheme 1). The nitration mechanism is not unclear at this moment although we suppose a coupling of cation radical **2a**$^{\bullet+}$ with nitrogen dioxide ($\bullet$NO$_2$) plays a key role. The crude product was purified by silica-gel column chromatography with hexane and toluene as the eluent. The structure of **2a** was unambiguously identified by ^{1}H-NMR, high-resolution LDI-TOF MS, and X-ray single-crystal structure analyses.

The ^{1}H-NMR spectrum of **2a** exhibited signals due to one β-proton of pyrrole and fifty protons of ethyl groups, and these signals appeared in the lower fields compared to those of **1a** due to an electron-withdrawing effect of the nitro group. Single crystals were fortunately obtained by vapor diffusion of methanol into the toluene solution of **2a**. The X-ray crystal structure analysis revealed the averaged bond length of the N-O moiety is 1.233(6) Å, the value of which is similar to that of nitrobenzene (1.226 Å) [30]. The molecules in the crystal adopted a slipped columnar structure with π–π stacking distances of 3.406 and 3.449 Å, calculated by the mean planes of 24 atoms forming hexaazacoronene. The values are almost the same as the sum of van der Waals (vdW) radii of carbon atoms (ca. 3.4 Å) (Figure 1A,B).

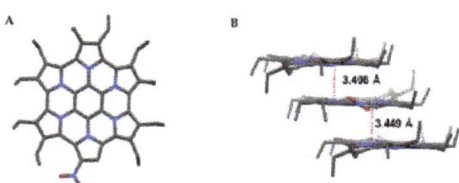

Figure 1. Crystal structures of **2a**. (**A**) Top view and (**B**) packing structures with the selected atomic distances.

To investigate the intramolecular charge transfer (ICT) character between the electron-rich HPHAC moiety and the nitro group, the absorption spectra of **1a** and **2a** were considered by the comparison with density functional theory (DFT) and the time-dependent density functional theory (TD-DFT) calculations of decamethylHPHACs (DMHPHACs **1b** and **2b**), where all ethyl groups of **1a** and **2a** were replaced by methyl groups for simplicity (Supplementary Information Sections 7, 8, 9, and 10). The cut-off of **2a** in a CH_2Cl_2 solution is bathochromically shifted by ca. 200 nm compared to that of **1a** (Figure 2). A characteristic broad band around 540–740 nm is ascribed to the mixed transitions of HOMO–LUMO and HOMO−1–LUMO, according to TD-DFT calculations (Supplementary Information Section 7). Figure 3 shows the molecular orbitals (MOs) of **1b** and **2b**. By the introduction of a nitro group on DMHPHAC **1b**, both HOMO and LUMO levels are lowered, and the decrease of the LUMO energy is larger than that of HOMO (ΔE = 0.86 for LUMO and 0.39 eV for HOMO). The LUMO is derived from the nitro group, mainly located at the nitro group, and slightly developed to the DMHPHAC skeleton. The LUMO+1 of **2b** is correlated to the LUMO of **1b**, and the LUMO+2 and LUMO+3 of **2b** are from degenerated LUMO+1 and LUMO+2 of **1b**. As a result, the HOMO–LUMO gap of nitroDMHPHAC **2b** is narrower than that of **1b**, which is attributable to the broad absorption band in the longest wavelength region of the ICT transition. In fact, positive solvatochromism was observed for **2a** in various solvents (Supplementary Information Section 4) because of strong stabilization of polarized LUMO of **2a** by polar solvents. NitroDEHPHAC **2a** showed little emission.

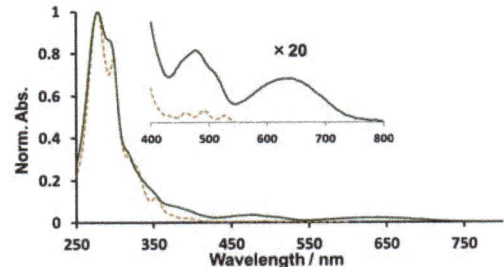

Figure 2. UV–VIS spectra of **1a** (orange) and **2a** (green) in CH_2Cl_2.

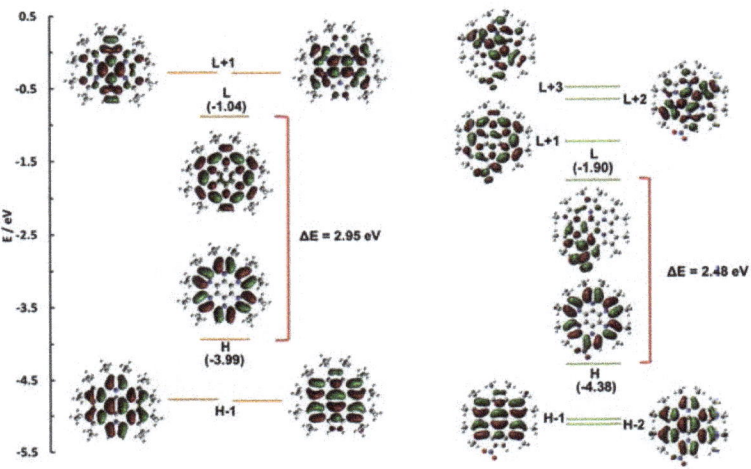

Figure 3. Molecular orbital diagrams of DMHPHACs **1b** and **2b** calculated at the B3LYP/6-31G(d,p) level.

Electrochemical properties of **2a** were examined by cyclic and differential pulse voltammograms (CV and DPV) (Figure 4). CV of **2a** showed two reversible oxidation waves at −0.20 and 0.07 V against a ferrocene/ferrocenium ion couple, values of which are low compared to those of **1a**. Thus, oxidative titration of **2a** was performed with pentachloroantimony(V) as monitoring the absorption spectra and resulted in the two-step changes with isosbestic points from neutral to radical cationic and then to dicationic species (Figure 5 and Supplementary Information Section 7). The peak maxima of radical cationic and dicationic species appeared in the visible to NIR region at 627 and 688 nm together with a broad band centered at ca. 1150 nm and at 784 and 884 nm with a broad band centered at ca. 1450 nm, respectively, which are characteristic for the previously reported the HPHAC radical cations and dications [15].

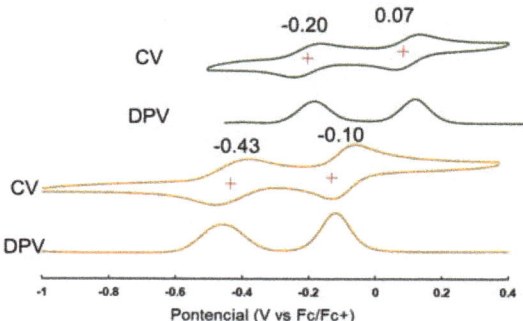

Figure 4. Cyclic and differential pulse voltammograms (CVs and DPVs) of DEHPHAC **1a** (orange) and **2a** (green). (Concentration: 0.1 M in CH_2Cl_2, supporting electrolyte: 0.1 M $TBAPF_6$).

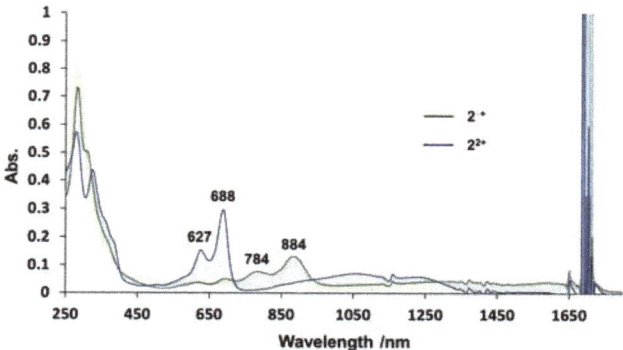

Figure 5. Stepwise oxidation of DEHPHAC **2a** with $SbCl_5$ in CH_2Cl_2 ([**2**] = 6.79 × 10^{-6} M).

For better understanding of the structures and electronic properties of oxidized species, the radical cation and dication of **2a** were aimed to be prepared by controlling the amount of an oxidant employed (Scheme 1). Both species were quantitatively obtained by the oxidation of proper amounts of silver(I) hexafluorophosphate as the oxidant, successfully isolated as hexafluorophosphate salts (isolated yields were not determined) and stable enough to handle under air in the solid states. Single crystals of **2a$^{•+}$[PF_6^-]** were obtained by slow vapor diffusion of hexane into the chlorobenzene solution and were subject to X-ray analysis. The crystal turned out to be composed of one molecule of **2a$^{•+}$[PF_6^-]** and one and a half molecules of chlorobenzene in the asymmetric unit, and the radical cation molecule of **2a$^{•+}$** was revealed to form a π-dimeric structure by an inversion center in the unit cell (Figure 6 and Supplementary Information Section 5). The closest atoms between the neighboring molecules in the π-dimer faced by the convex side are found in the benzene core, and the distance is 3.173 Å.

The distance of the mean planes of 24 hexaazacoronene atoms was 3.348 Å. These values are shorter than the sum of vdW radii of carbon atoms, strongly suggesting the π-dimer formation between two radical cations [31–34].

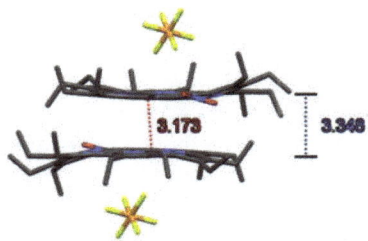

Figure 6. Crystal structure of ($2a^{\bullet+}[PF_6^-]$). The closest atomic (red) and plane (blue) distances are indicated.

The X-ray crystal structure of dication $2a^{2+}[PF_6^-]_2$ was also determined and illustrated in Supplementary Information Section 5. In general, the bond-length alternation (BLA) is important to discuss the nature of π-system. Table 1 shows the selected bond lengths of the pyrrole rings of neutral **2a**, radical cation $2a^{\bullet+}$, and dication $2a^{2+}$. By increasing the number of oxidation states, the averaged bonds of C_α-C_α and C_β-C_β become shorter, and the bond of C_α-C_β becomes longer, while that of C_α-N is nearly identical. These results indicate the delocalized radical and cation as observed for the previous system [15,16]. ^{1}H-NMR of $2a^{2+}$ clearly exhibits a downfield shift for the β-proton of pyrrole moiety (δ = 3.53 ppm) compared with that of neutral **2a** (Supplementary Information Section 6), demonstrating global aromaticity of $2a^{2+}$ as observed for $1a^{2+}$. The anisotropy of the induced current density (ACID) calculation for $2b^{2+}$ also supports global aromaticity (Supplementary Information Section 8). Notably, although the chemical shifts of β-protons of $2a^{2+}$ and $1a^{2+}$ are almost the same, ethyl protons of $2a^{2+}$ appear to be upfield-shifted (CH$_2$: Δδ ≈ 0.15 ppm) despite the existence of the electron-withdrawing nitro group. Based on the nucleus-independent chemical shift (NICS) calculation, global aromaticity of $2a^{2+}$ seems to be weaker than $1a^{2+}$ (Supplementary Information Section 8). Therefore, the slight upfield shifts of ethyl protons of $2a^{2+}$ would reflect the weaker global aromaticity, and the similar chemical shift values of β-proton of $2a^{2+}$ and $1a^{2+}$ observed would be due to cancellation of the conflicting effects of downfield (NO$_2$) and upfield (diminished global aromaticity).

Table 1. Selected bond lengths of **2a**, $2a^{\bullet+}[PF_6^-]$, and $2a^{2+}[PF_6^-]_2$ (the bonds in nitro-substituted pyrrole moiety were omitted due to the disorder) [Å] [1].

Bond	Neutral	Radical Cation	Dication
C_α-C_α	1.482(3)	1.463(8)	1.442(6)
C_β-C_β	1.435(3)	1.418(4)	1.396(5)
C_α-C_β	1.391(5)	1.407(7)	1.429(5)
C_α-N	1.394(4)	1.392(6)	1.389(7)

[1] Averaged values are shown with standard deviations calculated by the following Equation:

$$\{\sum(x_i - <x>)^2/(n-1)\}^{1/2} \tag{1}$$

3. Conclusions

The first example of nitration reaction of DEHPHAC **1a** giving nitroDEHPHAC **2a** was presented. The ICT character of **2a** from the electron-rich HPHAC core to the periferal nitro group was confirmed by diagnosis of the solvatochromiclly-shifting absorption centered at ca. 640 nm with TD-DFT calculations. Similar to the DEHPHAC **1a**, nitroDEHPHAC **2a** also showed stable redox properties. Crystallographic analysis of **2a** in not only neutral but radical cationic and dicationic states was performed to reveal a slipped columnar structure for the neutral molecule, π-dimeric structure for the radical cation, and anion-separated structure for the dication. These properties can be applied for supramolecular chemistry and organic electronics such as field effect transistor. Studies including the further derivatization and functionalization of HPHACs are currently underway in our laboratory.

4. Materials and Methods

^{1}H and ^{13}C-NMR spectra were recorded on JEOL (Akishima city, Japan) AL-400 at 400 MHz or Bruker Biospin (Osaka, Japan) AVANCE III at 500 MHz for ^{1}H and 100 MHz for ^{13}C with use of tetramethylsilane (0 ppm) or residual solvent for ^{1}H (7.26 ppm for CDCl$_3$) and ^{13}C (77.01 ppm for CDCl$_3$) signals, and if not otherwise noted, all spectra were measured in CDCl$_3$ at 298 K. Cyclic voltammetry (CV) measurements were performed on a CH Instruments-ALS612B electrochemical analyzer using a standard three-electrode cell consisting of Pt working electrodes, a Pt wire counter electrode, and an Ag/AgNO$_3$ reference electrode under a nitrogen atmosphere. The potentials were calibrated with ferrocene as an external standard. IR measurements were performed with Thermo Fischer Scientific (Tokyo, Japan) Nicolet iS5 FT-IR. Decomposition points were determined with Büchi M–565 and not corrected. Electronic spectra were recorded on a JASCO (Hachioji, Japan) V–570 UV–VIS/NIR spectrophotometer. High-resolution laser desorption/ionization time-of-flight mass spectra (HR LDI-TOF) were measured on JEOL (Akishima, Japan) JMS-S3000.

All reactions were carried out under nitrogen atmosphere. Thin-layer chromatography (TLC) analyses were carried out using silica gel 60 F$_{254}$ and aluminum oxide 60 F$_{254}$, neutral (Merck Millipore, Tokyo, Japan). Silica-gel chromatography was performed on Silica Gel 60 N (spherical, neutral) purchased from Kanto Chemical Co., Inc. (Tokyo, Japan) Alumina column chromatography was performed on activated alumina (about 200 mesh) purchased from FUJIFILM Wako Pure Chemical Corporation (Osaka, Japan). Dry solvents were purchased from Kanto Chemical Co., Inc. (dichloromethane).

4.1. Synthesis of N-[2,3,4,5,6-Pentakis(3, 4-Dimethylpyrrolyl)Phenyl]Pyrrole (DEHPB)

Sodium hydride dispersion in paraffin oil (ca. 60%, 380 mg) was weighed in a round flask, and the oil was removed by dry hexane. After drying, sodium hydride was weighed to be 207 mg (8.63 mmol), and the flask was sealed by a rubber septum. N-(2,3,4,5,6-Pentafluorophenyl)-1H-pyrrole [15] (180 mg, 0.773 mmol) in dry DMF (10 mL) was added to the flask by a syringe at 0 °C, then 3,4-diethylpyrrole (577 mg, 4.68 mmol) in dry DMF (5.0 mL) was added dropwise at the same temperature, the mixture was stirred, and then, the temperature was raised to 40 °C. After being stirred for 20 h, the reaction mixture was poured into water. The mixture was extracted with ethyl acetate. The organic extract was washed with water, dried over Na$_2$SO$_4$, and concentrated under a reduced pressure to give a crude product as a white solid. The crude product was purified by trituration with MeOH to give DEHPB as a white powder (468 mg, 0.624 mmol; 81%): ^{1}H-NMR (400 MHz) δ 0.93 (m, 30H), 2.21 (q, J = 7.4 Hz, 20H), 5.70 (s, 4H), 5.72 (s, 6H), 6.00 (m, 2H), 6.16 (m, 2H); ^{13}C-NMR (100 MHz) δ 14.80, 14.82, 14.84, 14.90, 14.93, 14.99, 18.49, 18.51, 110.08, 117.34, 117.45, 121.35, 126.39, 126.57, 126.58, 126.61, 132.13, 133.03, 133.41, 133.86. DI MS m/z 749 (M$^+$+1). IR (ATR) ν_{max} = 1494, 1488, 1122, 773, 723 cm^{-1} (Supplementary Information Sections 1–3).

4.2. Synthesis of 1,2,3,4,5,6,7,8,9,10-Decaethylhexapyrrolo [2 ,1,5-bc:2',1',5'-ef:2'',1'',5''-hi:2''',1''',5'''-kl: 2'''',1'''',5''''-no:2''''',1''''',5'''''-qr][2a,4a,6a,8a,10a,12a]Hexaazacoronene (DEHPHAC 1a)

To a CH$_2$Cl$_2$ (100 mL) solution of DEHPB (369 mg, 0.492 mmol) was added a CH$_3$NO$_2$ (5 mL) solution of FeCl$_3$ (2.8 g, 18 mmol) at room temperature. After being stirred for 3 h at 60 °C, the mixture was cooled to 0 °C, and then, hydrazine hydrate (5 mL) was added. The mixture was further stirred for 5 min at 0 °C. The mixture was extracted with toluene. The organic phase was washed with water, dried over Na$_2$SO$_4$, and concentrated under a reduced pressure. The residue was dissolved in a small amount of toluene and then passed through a short column of alumina. The eluate was concentrated to give a reddish brown solid. The solid was recrystallized from toluene/MeOH to give **1a** as orange crystals (250 mg, 0.359 mmol, 67%): ^{1}H-NMR (400 MHz, C$_6$D$_6$) δ 0.85 (m, 30H), 2.09 (q, J = 7.4 Hz, 20H), 5.66 (s, 2H); ^{13}C-NMR (100 MHz) δ 15.33, 16.98, 17.18, 17.87, 17.99, 18.05, 104.95 (2 kinds of C), 105.63, 106.25, 106.30, 106.60, 117.61, 117.98, 118.48, 118.77, 120.20, 122.01, 122.56, 122.72, 122.75, 122.99. HR-LDI-TOF MS: calcd. for C$_{50}$H$_{52}$N$_6^+$: 736.4253; found: 736.4287. IR (ATR) ν_{max} = 1630, 1448, 1371, 1360, 744 cm^{-1} (SI Sections 1–3).

4.3. Synthesis of nitroDEHPHAC 2a

Dichloromethane (50 mL) was added to a mixture of DEHPHAC **1a** (49 mg, 0.057 mmol) and AgNO$_2$ (100 mg, 0.65 mmol). The suspension was stirred vigorously at 20 °C for 1.5 h. After addition of hydrazine hydrate (ca. 1 mL), the solution was concentrated under a reduced pressure. A minimum solution of the residual solid in toluene was put on a silica-gel column filled with hexane, and then, the column was eluted with toluene. Green fractions were combined and then concentrated in vacuo to give **2a** (27 mg, 0.034 mmol, 51%) as a dark green solid: ^{1}H-NMR (400 MHz) δ 6.47 (s, 1H), 2.37–2.78 (m, 20H), 1.11–1.27 (q, 27H), 1.05 (t, 3H, ^3J = 7.3 Hz). ^{13}C-NMR (125 MHz) δ 133.75, 131.91, 126.19, 125.62, 125.00, 124.43, 124.10, 123.92, 123.79, 123.67, 122.87, 121.61, 120.73, 119.77, 119.35, 118.79, 108.25, 107.84, 106.60, 106.44, 106.29, 103.60, 102.02, 19.171, 18.14, 17.99, 17.657, 16.980, 16.87, 16.47, 15.31, 15.02. (The signals due to core benzene carbons were not found). HR-LDI-TOF MS: calcd. for C$_{50}$H$_{51}$N$_7$O$_2^+$: 782.4104; found: 782.4106. IR (ATR) ν_{max} = 1626, 1360, 1309, 1275, 1240 cm^{-1} (SI Sections 1–3).

Supplementary Materials: The following are available online at http://www.mdpi.com/1420-3049/25/11/2486/s1. Section 1: NMR spectra of **1a** and **2a**; Section 2: Mass spectra of DEHPB, **1a**, and **2a**; Section 3: IR spectra of DEHPB, **1a**, and **2a**; Section 4: Absorption spectra of **2a** in various solvents and dipole moments of **1b**, **2b**, and **2b**$^{2+}$; Section 5: X-ray structures of **2a**, **2a**$^{•+}$, and **2a**$^{2+}$; Section 6: 1H-NMR spectra of **2a**, **2a**$^{2+}$, and **1a**$^{2+}$ in CDCl3; Section 7: UV/Vis/NIR spectra of DEHPHACs **2a**, **2a**$^{•+}$ and **2a**$^{2+}$ with results of TD-DFT calculations of the corresponding DMHPHACs **2b**, **2b**$^{•+}$, and **2b**$^{2+}$; Section 8: ACID plots and NICS calculations of nitroDEHPHACs **2b** and **2b**$^{2+}$; Section 9: Bond lengths and ESP surfaces of **1b** and **2b** in the optimized structures; Section 10: Atomic coordinates of **2b**, **2b**$^{•+}$ and **2b**$^{2+}$ in the optimized structures.

Author Contributions: All authors contributed to the writing of the manuscript. Y.S. conducted all the experiments. All authors have read and agreed to the published version of the manuscript.

Funding: We appreciate JSPS KAKENHI Grant Numbers JP16K05698 and JP19K05422.

Conflicts of Interest: The authors declare no conflict of interest.

References

1. Yan, G.; Yang, M. Recent advances in the synthesis of aromatic nitro compounds. *Org. Biomol. Chem.* **2013**, *11*, 2554–2566. [CrossRef] [PubMed]
2. Ito, S.; Hiroto, S.; Lee, S.; Son, M.; Hisaki, I.; Yoshida, T.; Kim, D.; Kobayashi, N.; Shinokubo, H. Synthesis of Highly Twisted and Fully π-Conjugated Porphyrinic Oligomers. *J. Am. Chem. Soc.* **2015**, *137*, 142–145. [CrossRef] [PubMed]
3. Akita, M.; Hiroto, S.; Shinokubo, H. Oxidative Annulation of β-Aminoporphyrins into Pyrazine-Fused Diporphyrins. *Angew. Chem. Int. Ed.* **2012**, *51*, 2894–2897. [CrossRef] [PubMed]
4. Yokoi, H.; Wachi, N.; Hiroto, S.; Shinokubo, H. Oxidation of 2-Amino-Substituted BODIPYs Providing Pyrazine-Fused BODIPY Trimers. *Chem. Commun.* **2014**, *50*, 2715–2717. [CrossRef] [PubMed]
5. Goto, K.; Yamaguchi, R.; Hiroto, S.; Ueno, H.; Kawai, T.; Shinokubo, H. Intermolecular Oxidative Annulation of 2-Aminoanthracenes to Diazaacenes and Aza [7] helicenes. *Angew. Chem. Int. Ed.* **2012**, *51*, 10333–10336. [CrossRef]
6. Kosugi, Y.; Itoho, K.; Okazaki, H.; Yanai, T. Unexpected Formation of Dibenzo [a,h] phenazine in the Reaction Between 1-Naphthyliminodimagnesium Dibromide and 1-Nitronaphthalene. *J. Org. Chem.* **1995**, *60*, 5690–5692. [CrossRef]
7. Ueta, K.; Tanaka, T.; Osuka, A. Synthesis and Characterizations of *meso*-Nitrocorroles. *Chem. Lett.* **2018**, *47*, 916–919. [CrossRef]

8. Smith, K.M.; Barnett, G.H.; Evans, B.; Martynenko, Z. Novel meso-Substitution Reactions of Metalloporphyrins. *J. Am. Chem. Soc.* **1979**, *101*, 5953–5961. [CrossRef]
9. Saltsman, I.; Mahammed, A.; Goldberg, I.; Tkachenko, E.; Botoshansky, M.; Gross, Z. Selective Substitution of Corroles: Nitration, Hydroformylation, and Chlorosulfonation. *J. Am. Chem. Soc.* **2002**, *124*, 7411–7420. [CrossRef]
10. Stefanelli, M.; Mastroianni, M.; Nardis, S.; Licoccia, S.; Fronczek, F.R.; Smith, K.M.; Zhu, W.; Ou, Z.; Kadish, K.M.; Paolesse, R. Functionalization of Corroles: The Nitration Reaction. *Inorg. Chem.* **2007**, *46*, 10791–10799. [CrossRef]
11. Nishiyama, A.; Tanaka, Y.; Mori, S.; Furuta, H.; Shimizu, S. Oxidative Nitration Reaction of Antiaromatic 5,15-Dioxaporphyrin. *J. Porphyr. Phthalocyanines* **2020**, *24*, 355–361. [CrossRef]
12. Mikus, A.; Rosa, M.; Ostrowski, S. Isomers of β,β-Dinitro-5,10,15,20-Tetraphenylporphyrin Derivatives: Valuable Starting Materials for Further Transformations. *Molecules* **2019**, *24*, 838. [CrossRef]
13. Sun, B.; Ou, Z.; Yang, S.; Meng, D.; Lu, G.; Fang, Y.; Kadish, K.M. Synthesis and Electrochemistry of β-Pyrrole Nitro-Substituted Cobalt(ii) Porphyrins. The Effect of the NO_2 Group on Redox Potentials, the Electron Transfer Mechanism and Catalytic Reduction of Molecular Oxygen in Acidic Media. *Dalton Trans.* **2014**, *43*, 10809–10815. [CrossRef]
14. Yadav, P.; Kumar, R.; Saxena, A.; Butcher, R.J.; Sankar, M. β-Trisubstituted "Push–Pull" Porphyrins—Synthesis and Structural, Photophysical, and Electrochemical Redox Properties. *Eur. J. Inorg. Chem.* **2017**, *2017*, 3269–3274. [CrossRef]
15. Takase, M.; Enkelmann, V.; Sebastiani, D.; Baumgarten, M.; Müllen, K. Annularly fused hexapyrrolohexaazacoronenes: An extended π-system with multiple interior nitrogen atoms displays stable oxidation states. *Angew. Chem. Int. Ed.* **2007**, *46*, 5524–5527. [CrossRef] [PubMed]
16. Oki, K.; Takase, M.; Mori, S.; Shiotari, A.; Sugimoto, Y.; Ohara, K.; Okujima, T.; Uno, H. Synthesis, structures, and properties of core-expanded azacoronene analogue: A twisted π-system with two n-doped heptagons. *J. Am. Chem. Soc.* **2018**, *140*, 10430–10434. [CrossRef] [PubMed]
17. Navakouski, M.; Zhylitskaya, H.; Chmielewski, P.J.; Żyła-Karwowska, M.; Stępień, M. Electrophilic aromatic coupling of hexapyrrolylbenzenes. A mechanistic analysis. *J. Org. Chem.* **2020**, *85*, 187–194. [CrossRef]
18. Takase, M.; Narita, T.; Fujita, W.; Asano, M.S.; Nishinaga, T.; Benten, H.; Yoza, K.; Müllen, K. Pyrrole-fused azacoronene family: The influence of replacement with dialkoxybenzenes on the optical and electronic properties in neutral and oxidized states. *J. Am. Chem. Soc.* **2013**, *135*, 8031–8040. [CrossRef]
19. Gońka, E.; Chmielewski, P.J.; Lis, T.; Stępień, M. Expanded hexapyrrolohexaazacoronenes. Near-infrared absorbing chromophores with interrupted peripheral conjugation. *J. Am. Chem. Soc.* **2014**, *136*, 16399–16410. [CrossRef]
20. Żyła, M.; Gońka, E.; Chmielewski, P.J.; Cybińska, J.; Stępień, M. Synthesis of a Peripherally Conjugated 5-6-7 Nanographene. *Chem. Sci.* **2016**, *7*, 286–294. [CrossRef]
21. Żyła-Karwowska, M.; Zhylitskaya, H.; Cybińska, J.; Lis, T.; Chmielewski, P.J.; Stępień, M. An electron-deficient azacoronene obtained by radial π-extension. *Angew. Chem. Int. Ed.* **2016**, *55*, 14658–14662. [CrossRef] [PubMed]
22. Sasaki, Y.; Takase, M.; Okujima, T.; Mori, S.; Uno, H. Synthesis and Redox Properties of Pyrrole- and Azulene-Fused Azacoronene. *Org. Lett.* **2019**, *21*, 1900–1903. [CrossRef] [PubMed]
23. Uno, H.; Ishiwata, M.; Muramatsu, K.; Takase, M.; Mori, S.; Okujima, T. Oxidation Behavior of 1,3-Dihydrothieno[3,4-a] DEHPHAC. *Bull. Chem. Soc. Jpn.* **2019**, *92*, 973–981. [CrossRef]
24. Navakouski, M.; Zhylitskaya, H.; Chmielewski, P.J.; Lis, T.; Cybińska, J.; Stępień, M. Stereocontrolled synthesis of chiral heteroaromatic propellers with small optical bandgaps. *Angew. Chem. Int. Ed.* **2019**, *131*, 4983–4987. [CrossRef]
25. Oki, K.; Takase, M.; Mori, S.; Uno, H. Synthesis and isolation of antiaromatic expanded azacoronene via intramolecular vilsmeier-type reaction. *J. Am. Chem. Soc.* **2019**, *141*, 16255–16259. [CrossRef] [PubMed]
26. Moshniaha, L.; Żyła-Karwowska, M.; Chmielewski, P.J.; Lis, T.; Cybinska, J.; Gońka, E.; Oschwald, J.; Drewello, T.; Medina Rivero, S.; Casado, J.; et al. Aromatic Nanosandwich Obtained by σ-Dimerization of a Nanographenoid π-Radical. *J. Am. Chem. Soc.* **2020**, *142*, 3626–3635. [CrossRef]
27. Tokárová, Z.; Balogh, R.; Tisovský, P.; Hrnčariková, K.; Végh, D. Direct nucleophilic substitution of polyfluorobenzenes with pyrrole and 2,5-dimethylpyrrole. *J. Fluor. Chem.* **2017**, *204*, 59–64. [CrossRef]

28. Ooi, S.; Yoneda, T.; Tanaka, T.; Osuka, A. meso-Free Corroles: Syntheses, Structures, Properties, and Chemical Reactivities. *Chem. Eur. J.* **2015**, *21*, 7772–7779. [CrossRef]
29. Osuka, A.; Shimidzu, H. meso, meso-Linked Porphyrin Arrays. *Angew. Chem. Int. Ed. Engl.* **1997**, *36*, 135–137. [CrossRef]
30. Boese, R.; Biäser, D.; Nussbaumer, M.; Krygowski, T.M. Low temperature crystal and molecular structure of nitrobenzene. *Struct. Chem.* **1992**, *3*, 363–368. [CrossRef]
31. Kertesz, M. Pancake Bonding: An Unusual Pi-Stacking Interaction. *Chem. Eur. J.* **2019**, *25*, 400–416. [CrossRef]
32. Tateno, M.; Takase, M.; Iyoda, M.; Komatsu, K.; Nishinaga, T. Steric Control in the π-Dimerization of Oligothiophene Radical Cations Annelated with Bicyclo [2.2.2] octane Units. *Chem. Eur. J.* **2013**, *19*, 5457–5467. [CrossRef] [PubMed]
33. Morita, Y.; Suzuki, S.; Fukui, K.; Nakagawa, S.; Kitagawa, H.; Kishida, H.; Okamoto, H.; Naito, A.; Sekine, A.; Ohashi, Y.; et al. Thermochromism in an organic crystal based on the coexistence of σ- and π-dimers. *Nat. Mater.* **2008**, *7*, 48–51. [CrossRef] [PubMed]
34. Yokoi, H.; Hiroto, S.; Shinokubo, H. Reversible σ-Bond Formation in Bowl-Shaped π-Radical Cations: The Effects of Curved and Planar Structures. *J. Am. Chem. Soc.* **2018**, *140*, 4649–4655. [CrossRef] [PubMed]

Sample Availability: Samples of the compounds DEHPB, **1a**, and **2a** are available from the authors.

© 2020 by the authors. Licensee MDPI, Basel, Switzerland. This article is an open access article distributed under the terms and conditions of the Creative Commons Attribution (CC BY) license (http://creativecommons.org/licenses/by/4.0/).

Review

Nitro-Perylenediimide: An Emerging Building Block for the Synthesis of Functional Organic Materials

Lou Rocard, Antoine Goujon and Piétrick Hudhomme *

Laboratoire MOLTECH-Anjou, UMR CNRS 6200, UNIV Angers, SFR MATRIX, 2 Bd Lavoisier, Angers CEDEX 49045, France; lou.rocard@univ-angers.fr (L.R.); antoine.goujon@univ-angers.fr (A.G.)
* Correspondence: pietrick.hudhomme@univ-angers.fr; Tel.: +33-2-4173-5094

Academic Editor: Nagatoshi Nishiwaki
Received: 5 March 2020; Accepted: 17 March 2020; Published: 19 March 2020

Abstract: Perylenediimide (PDI) is one of the most important classes of dyes and is intensively explored in the field of functional organic materials. The functionalization of this electron-deficient aromatic core is well-known to tune the outstanding optoelectronic properties of PDI derivatives. In this respect, the functionalization has been mostly addressed in bay-positions to halogenated derivatives through nucleophilic substitutions or metal-catalyzed coupling reactions. Being aware of the synthetic difficulties of obtaining the key intermediate 1-bromoPDI, we will present as an alternative in this review the potential of 1-nitroPDI: a powerful building block to access a large variety of PDI-based materials.

Keywords: perylenediimide; nitro group; organic materials

1. Introduction

Perylenediimide (PDI) derivatives, discovered by Kardos in 1913 [1], have been extensively studied and exploited in many fields. First, taking advantage of their outstanding high chemical, thermal and photochemical stability [2], they have been initially used as red dyes and pigments [3,4], in paints, lacquers and reprographic processes [5], This application results from the strong visible-light absorbing capabilities of PDI, whose UV-vis spectrum displays three characteristic bands spanning from 430 to 550 nm with high extinction coefficient (up to $\approx 10^5$ M^{-1} cm^{-1}). Those optical properties are the consequence of the strong conjugation within its molecular structure corresponding to an electron-rich polyaromatic core, the so-called perylene, substituted with two electron-withdrawing imide groups at the 3,4 and 9-10-*peri*-positions. Besides, PDI derivatives display very high quantum yields of fluorescence (close to 1), which enables their development as fluorescent dyes [6], near-IR dyes [7], molecular switches [8] and dye lasers [9]. In addition to their excellent photophysical properties, the electron-poor character of PDI is highlighted by a first reduction potential around −1 eV vs. the ferrocenium/ferrocene (Fc$^+$/Fc) couple. These characteristics make PDI a strong electron acceptor, with low lying LUMOs and high electron mobility; it is recognized among the best n-type semiconductors available to date and provides strong prospects for investigations in organic solar cells (OSCs) as a good alternative to fullerene [10–13]. Owing to those particular opto- and electrochemical characteristics, PDIs promote photoinduced electron and/or energy transfer processes and were intensively exploited as building blocks to construct light-harvesting arrays and artificial photosynthetic systems [14–22]. Finally, besides the evident applications in electronic devices, the interest for those properties is growing in medicine, such as for photodynamic therapy [23], as PDIs have shown biocompatibility [24] when correctly decorated with water-soluble groups.

For all those applications relative to absorption, emission and electron-accepting capabilities, the fine and desired tailoring of the photoredox properties through the tuning of the HOMO-LUMO band gap of PDI is required. Among all the synthetic tools to do so, the most popular methods consist in (1) modifying the size of the conjugated system; (2) changing the planarity of the molecule to affect the π-π interactions by introducing bulky substituents or promoting annulations; (3) inserting electro-donating or withdrawing groups; (4) replacing a carbon atom by a isoelectronic heteroatom (doping method) [25–27]. All those strategies have been widely applied to PDI derivatives, mostly through the functionalization of the perylene core at the ortho or bay positions, as modifications at the imide position barely influence the photoredox properties but mostly affect the solubility of the material [28,29]. Historically, the functionalization of the perylene core has been achieved via the formation of key halogenated-PDIs intermediates in the positions 1,12 and 6,7 (bay region); in particular, 1,6,7,12-tetrachlorinated or 1-mono and 1,6/1,7-dibrominated PDIs. In this review, we will present the emergence of nitro-substituted PDIs as new key intermediates for modifying the molecular structure of the PDI core and efficiently tuning its HOMO-LUMO band gap (Figure 1). A particular attention will be given to the monofunctionalization of PDI at the bay position (position 1).

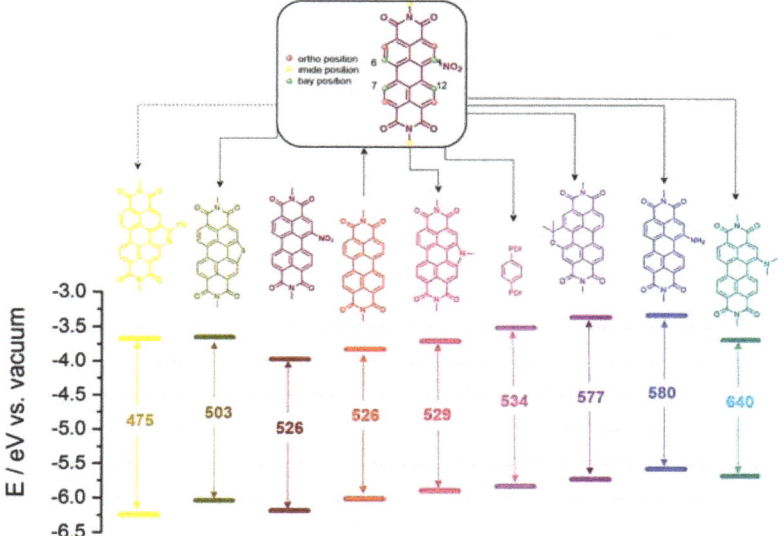

Figure 1. 1-NitroPDI, a key intermediate for reaching various bay-substituted-PDI structures with distinct photoredox properties. Frontier orbital energy (HOMO-LUMO) levels of PDIs. Wavelengths (nm) of maximal absorption of light are shown. The data were collected from references [30–35].

2. Nitration vs. Bromination Conditions of Perylenediimide Derivatives

Nitration or halogenation, and more specifically chlorination or bromination, on the perylene core through electrophilic substitution (S_EAr) affords key building blocks for further functionalization of PDI derivatives. Such a reaction enables the decoration of the bay positions with respect to Holleman rules, according to the electron withdrawing character of the diimide groups. Historically, the tetra-chlorination of PDIs has enabled the emergence of wide applications by increasing solubility in organic solvents with a significant twisting of the perylene core and the possibility to tune their electronic properties [36]. In this review, we are focusing on the bis, and essentially, the mono-bay-functionalization of PDIs, which limit the twisting of the perylene. To do so, the most popular method is the introduction of bromine atom(s), which can subsequently be replaced through nucleophilic substitution or pallado-catalyzed organometallic coupling reactions. We are herein highlighting that the (mono) nitration of PDI should be considered as an excellent alternative to the (mono)bromination.

For the mono and bis bromination reactions, numerous synthetic strategies have been reported [29]. Nevertheless, the most used pathway using the milder conditions is shown in Scheme 1. Typically, the reactions occur in a chlorinated solvent, in the presence of a large excess (>50 equiv) of bromine at room temperature for more than 2 days to reach 1-bromoPDI **3**, or under reflux for more than 1 day to obtain a mixture of regioisomers 1,6 and 1,7-dibromoPDI **2** in ratio ≈1:3 [37]. The mono-bromination conditions afforded a complicated mixture of dibromo (25%), monobromo (50%) and unreacted PDI requiring fastidious purification by chromatography. Moreover, practically, the quenching of large excess of bromine is a critical step and limits the large-scale use of this essential precursor of PDI-based materials.

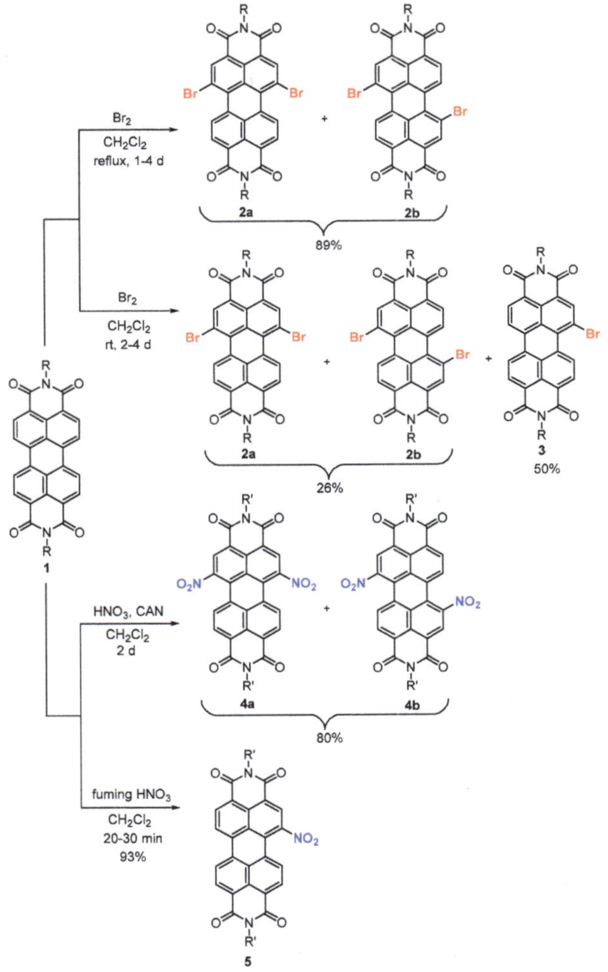

Scheme 1. Mono and bis-nitration vs. mono and bis-bromination of PDI. R = 1-ethylpropyl; R' = cyclohexyl.

The mononitration of PDI **1** is more selective, owing to the electron-withdrawing inductive and mesomeric effects of the nitro group, which sufficiently deactivates the PDI core towards the second electrophilic substitution. The reaction was initially performed by Langhals and colleagues, who attained 98% yield using a solution of N_2O_4 gas (prepared by strong heating of $Pb(NO_3)_2$) and methanesulfonic acid (CH_3SO_3H) as the catalyst in CH_2Cl_2 [38,39], and this was achieved more easily using nitric acid in the presence of cerium (IV) ammonium nitrate (CAN) in CH_2Cl_2 with nearly quantitative yields [30,31,40–46]. The kinetics of the reaction were reported to be higher using a mixture of HNO_3 and H_2SO_4 [47–49]. Later, it was shown that the nitration reaction could be carried out using an excess of fuming nitric acid (≈25 equiv), wherein the addition of CAN was not improving the yield or favoring the kinetic of the reaction [32,50]. allowing the preparation of mononitroPDI **5** in 93% yield (at room temperature and in short time) in a multigram scale without the need of purification by silica gel chromatography (Scheme 1) [51]. Under the same conditions, and only by carrying out the reaction for a longer time (6 h), the bis-nitration can occur to afford the two regioisomers (1,6)-and (1,7)-dinitroPDI **4** with no, or almost no selectivity [52]. Notably, when the reaction was described in the presence of CAN for 48 h, the authors claimed a selectivity in favor of the regioisomer (1,7) **4b** (3:1) and the possibility of separating the two regioisomers by HPLC or repetitive crystallizations [40,53].

Hence, the functionalization of PDI and in particular its mono-functionalization is far more efficient in terms of yield, time reaction, purification, atom economy, etc., through its nitration than its bromination.

3. Reactivity of 1-NitroPerylenediimide

3.1. Nucleophilic Substitution of the Nitro Group

The fine and controlled tuning of the electrochemical and optical properties of PDIs is of great interest for engineering fluorophores, color pigments, acceptors in organic transistors and solar cells, etc. To respond to this demand, the synthetic strategy is to attach electron-donating or electron-withdrawing groups on the perylene core. The functionalization with amine and alcohol through nucleophilic substitution is the most common approach. The amino group on the bay position, for instance, drastically changes the optoelectronic properties of PDI, with a long-wavelength charge transfer band, owing to the HOMO located on the amine and the LUMO on the PDI. This strategy has been widely explored using halogenated PDI; in particular, bromo-PDI [29]. As an example, the replacement of bromine atoms in a mixture of 1,6 and 1,7-dibromoPDIs **2** by pyrrolidine moieties was carried out in 80% yield heating the reaction in neat pyrrolidine at 50 °C for 18 h [54].

Thanks to the strong electron-withdrawing character of nitro group, nucleophilic substitution reactions can be easily achieved from 1-nitroPDI **5** (Scheme 2). X. Kong et al. described smooth substitutions of 1-nitroPDI with phenol derivatives, ethanol, propanethiol and pyrrolidine as nucleophiles [55]. When stirring at room temperature with 4-tertbutylphenol in the presence of K_2CO_3 and a catalytic amount of KI in NMP, the reaction afforded the desired product **6a** in 90% yield. The reaction can occur in DMF or $CHCl_3$ in lower yields (less than 65%). Similar conditions were also described to replace the nitro group by electron poor phenoxy and aliphatic ether and thioether in $CHCl_3$ or DMF [48,49], which required higher temperature when using aliphatic thiol or alcohol. Remarkably, stirring 1-nitroPDI in neat pyrrolidine at 0 °C already promoted the nucleophilic substitution to give **9** in 30% yield, which highlights the high reactivity of the nitro group toward S_NAr [55]. Nevertheless, when the reaction was carried out above 25 °C, the major product surprisingly formed was 1,6-disubstituted PDI **10** without any further discussion from the authors justifying the presence of this side product.

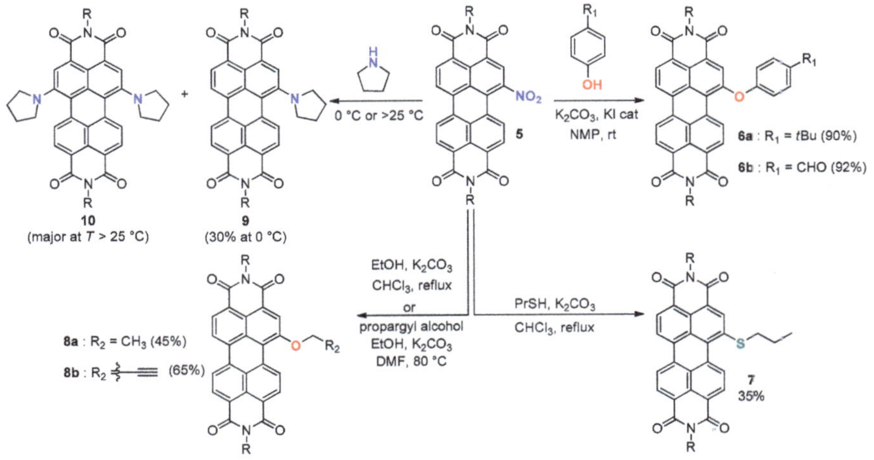

Scheme 2. Nucleophilic substitutions from 1-nitroPDI **5**. R = cyclohexyl.

3.2. Access to Core-Extended Annulated PDI

The incorporation of annulated-heteroatoms such as S [56], Se [46], or N [50], in the bay positions of the PDI core is recognized to be one of the most effective method to decrease its electron affinity as well as to reduce the intermolecular aggregation between PDI units [57]. In this respect, important research effort has focused on the synthesis of heteroatom-annulated PDIs for the elaboration of high performance organic semiconductors and organic solar cells. To the best of our knowledge, 1-nitroPDI **5** is the most convenient starting material to reach those structures.

Pioneering results were obtained by H. Langhals et al. to reach five-membered rings S- and N-annulated PDIs [38]. The reaction between nitroPDI **5** in neat triethylphosphite, known as Cadogan cylization, yielded to a mixture of carbazole fused PDI **11** and phosphorylated PDI **12** (Scheme 3). The reducing agent promoting the formation of the nitrene intermediate can indeed give rise to a competitive nucleophilic substitution of the nitro group. The protocol was later improved by G. C. Welch and colleagues, who replaced the phosphorous ester by triphenylphosphine in DMF affording desired product **11a** in 67% yield [50]. This yield was considerably increased (91%) when using microwave irradiation in o-dichlorobenzene at 180 °C for 2 h [58]. An interesting alternative to this method, which does not require high temperature or generate phosphine oxide, was found by serendipity in our group. Adding a small excess of sodium azide to nitro-PDI in a THF/DMF mixture at room temperature afforded the N-annulated PDI **11b** in 76% yield [33]. We assume that a nucleophilic substitution first occurs followed by the formation of a nitrene intermediate, which spontaneously led to the carbazole ring. This N-containing ring allows the introduction of an extra solubilizing chain using an alkyl halide, or the functionalization with another chromophoric unit to reach a dyad [59]. or the dimerization of N-annulated PDI with a linker via a Buchwald-Hartwig coupling [58].

Moreover, in the original work of H. Langhals, the authors also described the use of sulfur on nitroPDI **5** in DMF or NMP, leading to a mixture of two sulfur containing heterocycles **14** and **15** in different ratio depending on the solvent and the temperature. Notably, at a high temperature in DMF, only thiophene-annulated PDI **14** was formed in 71% yield [38]. L. Chen et al. reported the formation of derivatives **14** and **15** in 34% and 42% yields respectively, by heating the reaction in DMF at 120 °C [60]. On the other hand, bubbling O_2 or adding selenium powder into 1-nitroPDI **5** in NMP at 180–190 °C led to O-annulated PDI **16** [61]. or selenophene-heterocycle **17** in 30% and 80% yields, respectively [46]. Those syntheses were also applied to the bay-annulation of the perylene tetraester by S. Achalkumar and colleagues [34]. Besides, the authors nicely reported the effect of the heteroatom on the molecular self-assembly.

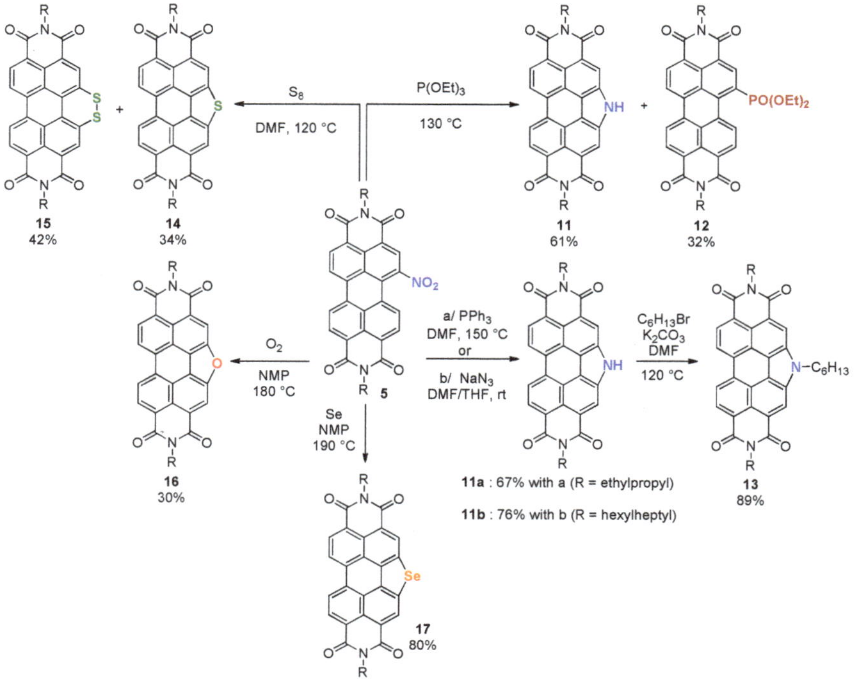

Scheme 3. Mono-heteroatom-annulation from 1-nitroPDI **5**. R = cyclohexyl.

The possibility of introducing two alkyl chains to enhance the solubility of heteroatom-fused PDIs was explored with the synthesis of pyran-annulated PDI compounds (Scheme 4) [32]. Monopyran-fused PDIs **18** were prepared from 1-nitroPDI via a one-pot nucleophilic substitution/cyclization sequence using 2-nitro-propane or diethyl malonate in NMP at room temperature in 85% and 79%, respectively. The nitration of those derivatives led selectively and almost quantitatively to 7-nitro-pyran-fused PDIs **19**, possibly owing to the electron donating character of the pyran. By subsequent cyclization under the same conditions, bispyran-fused PDIs **20** were obtained.

Following the same sequential synthesis, S-annulated PDIs **14** and **15** were nitrated and bis-bridged, affording unsymmetrical S-PDI-2S **21** in both cases in 76% yield over two steps [60]. In the same work, a one-pot synthesis of the same product was also reported from a mixture of 1,6 and 1,7-dinitro-PDIs **4** in 80% yield. In addition, the S and O-doping strategies were mixed, and in this respect, 7-nitro-pyran-fused PDI **19b** was S-annulated in the presence of sulfur powder yielding to the mixture of two sulfur- and oxygen-containing heterocycle PDIs **22** and **23**. Notably, the two derivatives display very different optical properties, as depicted in the pictures in Scheme 4 [62].

Finally, the bis-sulfur-annulation was applied to a fused-PDI dimer **24** through nitration of the parent dimer (Scheme 5) [63]. The reaction with sulfur powder in refluxing NMP afforded a mixture of multisulfur-fused-PDIs **25**, **26** and **27**.

The three products display different optoelectronic properties, and their performances as non-fullerene acceptors (NFAs) in organic bulk heterojunction (BHJ) were demonstrated with power conversion efficiency (PCE) up to 6.9% and high fill factors (>60%).

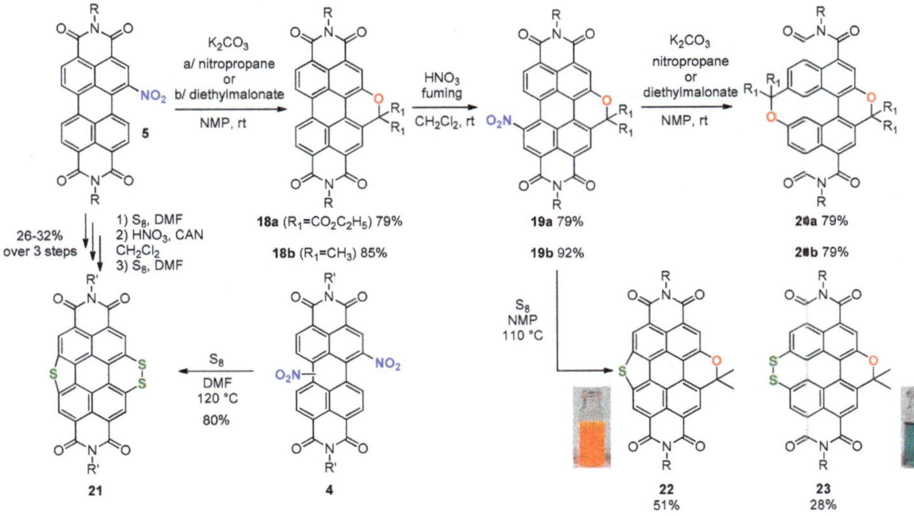

Scheme 4. Bis-heteroatom-annulation from nitroPDI derivatives. R = cyclohexyl, R′ = 1-ethylpropyl. Pictures adapted with permission from reference 62. Copyright (2018) John Wiley and Sons.

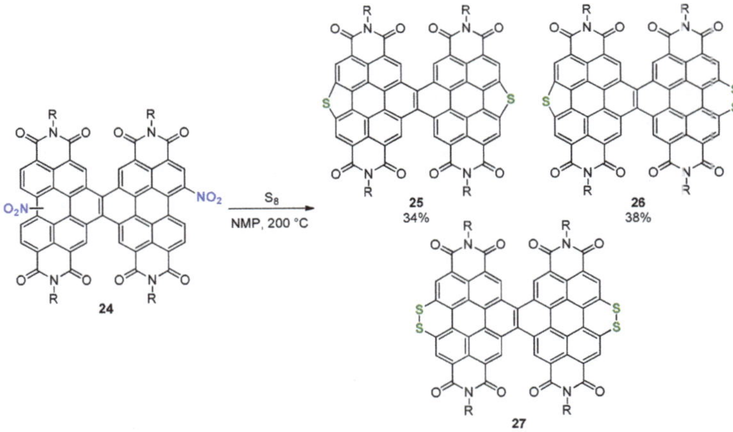

Scheme 5. S-annulated fused PDIs. R = 1-pentylhexyl.

3.3. Palladium-Catalyzed Cross-Coupling Reactions

The bay-decoration of PDI through the formation of C-C bond has been widely employed in organic electronics, such as for the elaboration of OSCs using donor-acceptor (D-A) systems or PDI-based NFAs. The synthetic methodology usually followed is the bromination of PDI in bay position and subsequent Pd-catalyzed cross-coupling reaction. However, as already discussed, the monobromination of PDI suffers from many disadvantages compared to the mononitration (see Section 2).

Recently, the use of nitroarenes as electrophilic partners in Pd-catalyzed couplings for the creation of C-heteroatoms and C-C bonds has been recognized [64]. In particular, the Suzuki–Miyaura coupling (SMC) with nitroarenes has been demonstrated in 2017 by Y. Nakao and S. Sakaki [65,56].

The first example of a SMC reaction using an electron-deficient arene system such as PDI **5** bearing a nitro group as the electrophilic coupling partner was reported in our group [51]. This reaction uses a straightforward procedure with 4-formyl or 3-formyl phenylboronic acid in the presence of Pd(PPh$_3$)$_4$ and K$_3$PO$_4$ as a base in refluxing THF, affording **28** and **29** in 85% and 81% yields, respectively (Scheme 6). Subsequent 1,3-dipolar cycloaddition using C$_{60}$ was carried out on 1-(3-formylphenyl)PDI **29** derivative. The versatility of this original SMC reaction was demonstrated by replacing the phenyl boronic acid substituted with the electron-withdrawing formyl group by the (4-diphenylamino)phenyl electron-donating group. In 2019, J. K. Kallistsis and co-workers applied the same conditions to functionalize the PDI with a protected phenoxy and a styryl moiety [67]. Those anchoring groups were subsequently used to attach withdrawing quinoline derivatives through Heck coupling or nucleophilic substitution after the alcohol deprotection.

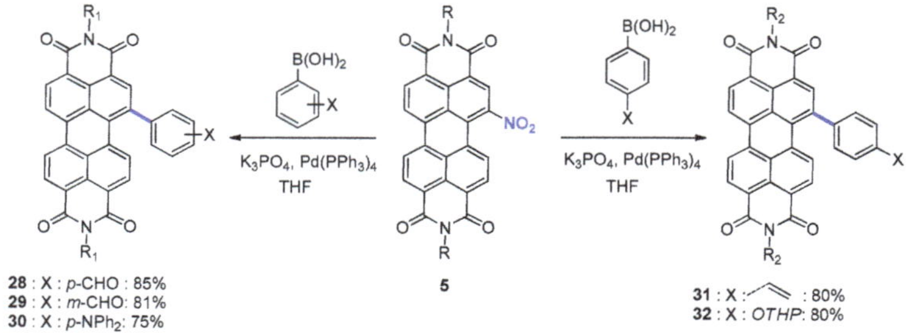

Scheme 6. Suzuki–Miyaura coupling reactions from 1-nitroPDI **5**. R = cyclohexyl (R$_1$) or 2-ethylhexyl (R$_2$).

Besides, over the last decade, PDI-based multimers have received as much attention as NFAs in organic photovoltaics (OPV) owing to their three-dimensional structure preventing the formation of aggregates and in this respect enhancing the performance of OSCs. In our group, we developed conditions of multimerization through poly-SMC using an aromatic linker with multiple boronic acids or esters and 1-nitroPDI or 1-bromoPDI as electrophilic partners [33]. The described methodology employed Pd(PPh$_3$)$_4$ as the Pd(0) source with K$_2$CO$_3$ in a dioxane/water mixture under microwave irradiations, affording various dimers, trimers and tetramers (Scheme 7 and Figure 2). In general, the yields are comparable, starting from nitro **5** and bromo **3** PDI derivatives, and are notably good for the phenylene (Ph) series (over 90% per coupling from 1-nitroPDI).

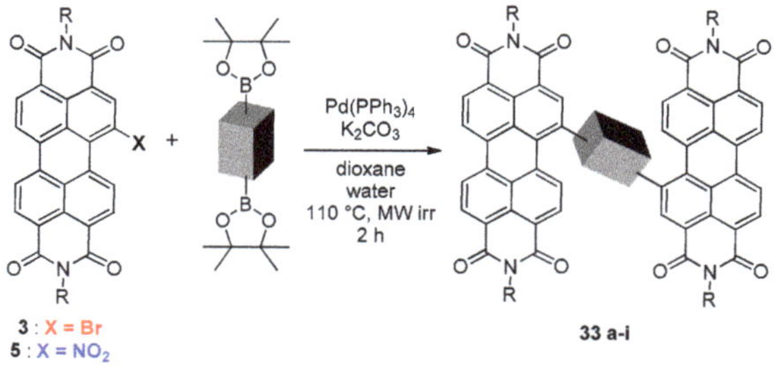

Scheme 7. PDI multimerization through SMC from 1-nitroPDI and 1-bromoPDI. R = 1-hexylheptyl.

However, when the multimerization conditions were applied to enriched *N*-annulated bromo- or nitro-PDIs **34** and **35** in the presence of bis-boronate phenylene, significantly different reaction behaviors were observed (Scheme 8). The reaction carried out with bromo derivative **34** was clean and complete, whereas when applied to nitro-PDI **35**, the product **36** was obtained in 24% yield with 40% of starting material **35** recovered. This result coupled with theoretical calculations suggests that the limiting step of the coupling, the oxidative addition, is favored with the bromo derivative. Nevertheless, the electron-deficiency of the PDI core seems to promote this step, as no significant difference was noticed between the two electrophilic derivatives **3** and **5** during the parent PDI multimerization. In this respect, other Pd-catalyzed cross-couplings using nitro-PDI can be expected to be soon developed for the preparation of unprecedented PDI-based materials.

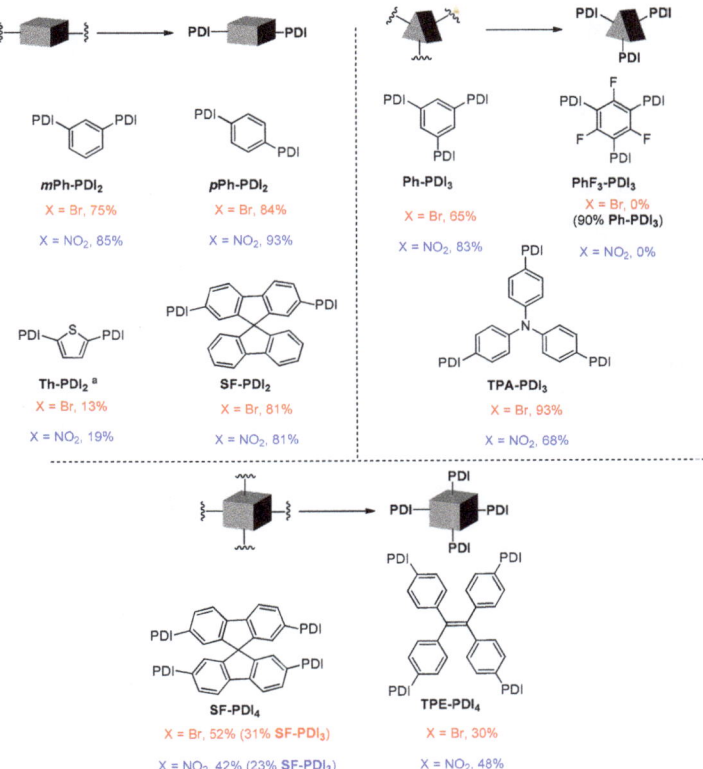

Figure 2. Schematic structures of PDI-based multimers obtained through SMC from 1-nitroPDI **5** and 1-bromoPDI **3**. Figure adapted with permission from reference 57. Copyright (2019) John Wiley and Sons.

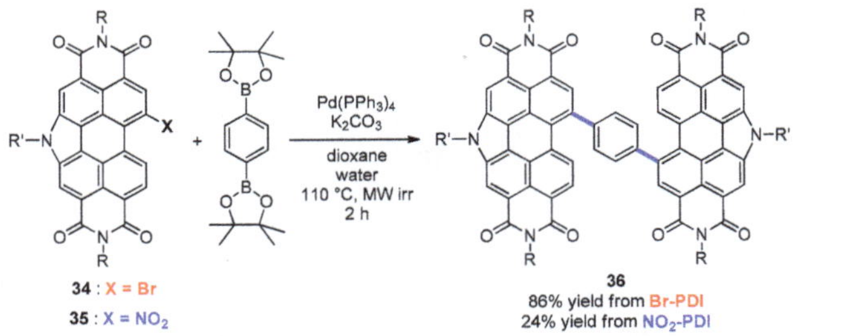

Scheme 8. N-annulated PDI multimerization through SMC from 1-nitroPDI and 1-bromoPDI. R = 1-hexylheptyl; R' = octyl.

4. Reduction of the Nitro Group and Functionalization of AminoPDI

Nitroarenes are classically used as key intermediates for the synthesis of aniline derivatives obtained by reduction. Such a transformation of 1-nitroPDI into 1-aminoPDI can be performed using different procedures, affording another interesting building block to design original materials for applications in organic electronics. This transformation and subsequent developments will be presented in this section.

4.1. Preparation 1-aminoPDI from 1-nitroPDI

Langhals and colleagues reported first the reduction of 1-nitroPDI **5** into 1-aminoPDI **37** using iron powder and hydrochloric acid in either ethanol or THF in 81% yield, along with an hydrogenation procedure catalyzed by Pd on carbon with triethylammonium formate as an hydrogen source (Scheme 9) [38]. However, the reactant of choice to perform the reduction of the nitro group into the corresponding amino group in PDI series remains tin (II) chloride dihydrate ($SnCl_2.2H_2O$) in the refluxing of THF, affording the amine with yields of around 80% [30,31,41,42,44,45,47,68]. Alternatively, reduction involving metallic Zn and acetic acid in THF led to 1-aminoPDI in 95% yield [42], whereas metallic iron in the presence of HCl in THF gave the derivative in 70% yield [39]. Another common protocol for reducing nitroarenes to arylamines is the catalytic hydrogenation on Pd/C using hydrazine as an hydrogen source in DME at 80°C, which quantitatively afforded 1-aminoPDI derivative **37** [69]. This method could also be adapted to the direct use of hydrogen gas in THF at room temperature, but Q. Zhang and colleagues noted that N,N'-(dicylohexyl)1-aminoPDI had to be directly engaged in the following steps without further purification, as instability in air was observed [70].

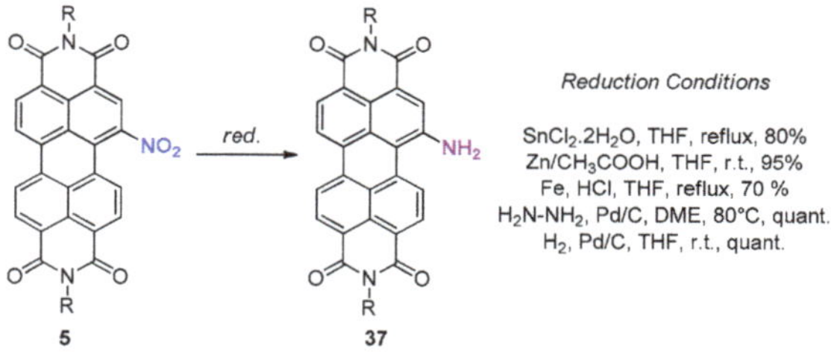

Scheme 9. Reduction of 1-nitroPDI **5**.

1,6-and 1,7-dinitroPDIs could also be reduced into diaminoPDI using similar procedures. $SnCl_2$ was used to transform 1,7-dinitroPDI into diamino derivative in 82% yield [27].

AminoPDI derivatives display a broad charge-transfer band in absorption accompanied by partial or total quenching of their fluorescence and show strong solvatochromism. They are key intermediates for the synthesis of alkyl-aminoPDI derivatives, metal complexes and the preparation of bay-extended PDI architectures, providing sensors and NIR-emitters for examples.

4.2. Reactivity of 1-AminoPDI

The alkylation of the amino group was reported using NaH as a base followed by the addition of alkyl iodide providing dialkylated aminoPDI **38a** in an 85% yield (Scheme 10) [30]. Acylation of the amino group was also carried out in only 33% yield using acetyl chloride and pyridine in THF [42], or 2-pyridyl acyl chloride in the presence of Et_3N to afford fluorescent and colorimetric sensors **39** and **40** for Cu^{2+} and F^- ions—in which charge transfers do not occur. In the presence of F^- anions, deprotonation of the amide seems to take place, whereas in the presence of Cu^{2+}, a dimeric complex is formed. In both cases, this leads to a progressive extinction of the fluorescence and red-shift of the absorption maximum [43].

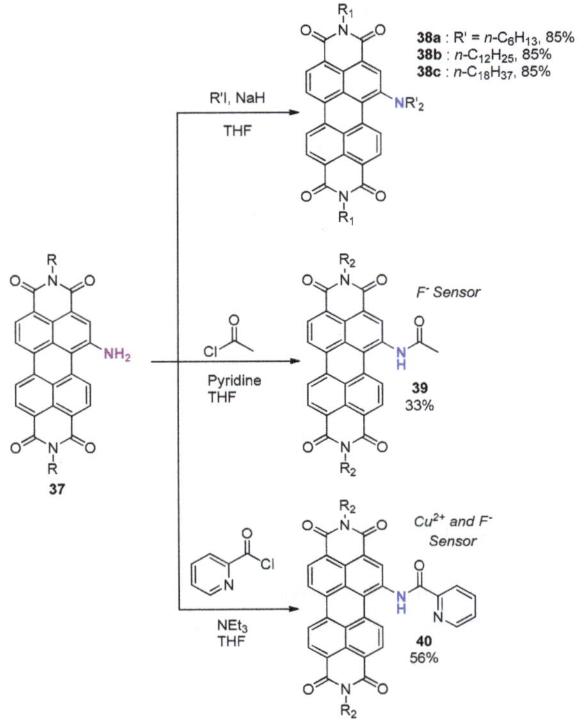

Scheme 10. Alkylation and acylation of aminoPDI. R_1 = 1-ethylhexyl, R_2 = cyclohexyl.

Treatment with bis(trichloromethyl) carbonate (BTC) in the presence of triethylamine in THF followed by the reaction with allyl alcohol led to the formation of the fluorescent mono-carboxylated amino intermediate (Scheme 11). The latter was further alkylated with allyl bromide to afford a colorimetric and fluorescent sensor for Pd^{2+} detection in 46% yield [71]. In the presence of Pd^0, decarboxylation of compound **41** led to the formation of a non-fluorescent species, which also shows its absorption maximum shifting toward longer wavelength.

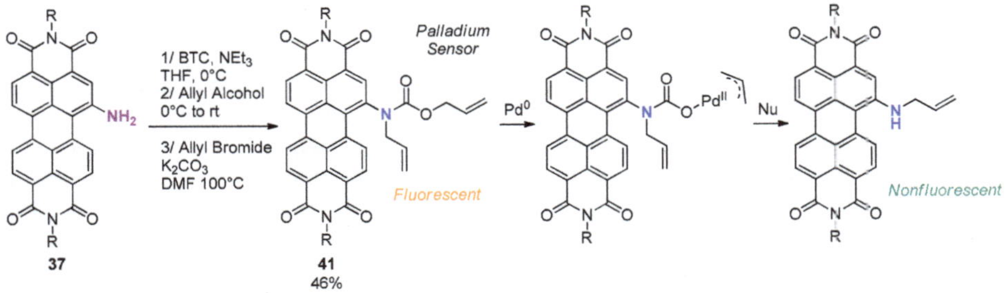

Scheme 11. Preparation of a colorimetric and fluorescent Pd^0 probe. R = 2-ethylhexyl.

Isocyanide derivative **42** also became a target of choice for its ability to form metal complexes, and for being a precursor of carbene complexes. The synthesis of 1-isocyanidePDI required the formylation of 1-aminoPDI **37** with formic acid in 95% yield followed by its dehydration using triphosgene and triethylamine in CH_2Cl_2, leading to desired compound **42** in 57% yield (Scheme 12) [39]. Corresponding gold carbene complexes **43** were prepared by coordination of the isocyanide group with Au(I) precursors in very mild conditions. Multi-nuclear architectures were also prepared and some of them were processed as stable Langmuir–Blodgett films.

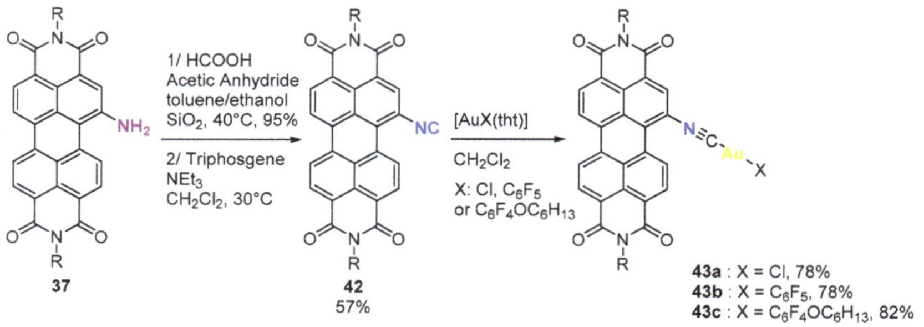

Scheme 12. Preparation of 1-isocyanidePDI **42** and its complexation with Au(I). R = 1-ethylpropyl.

Besides, a peculiar azoborine-annulated PDI architecture was reported by Q. Zhang and coll as a highly selective fluoride anion sensor (Scheme 13). This first B-N annulation was performed in 81% yield by reacting 1-aminoPDI with dichlorophenylborane in refluxing toluene and using triethylamine as a base [70]. Compound **44** is an efficient F$^-$ sensor, showing extinction of its fluorescence and red-shifted absorption maximum with increasing concentration of anion. This boron fused PDI was also used as an emitter in OLED devices, emitting in the red region.

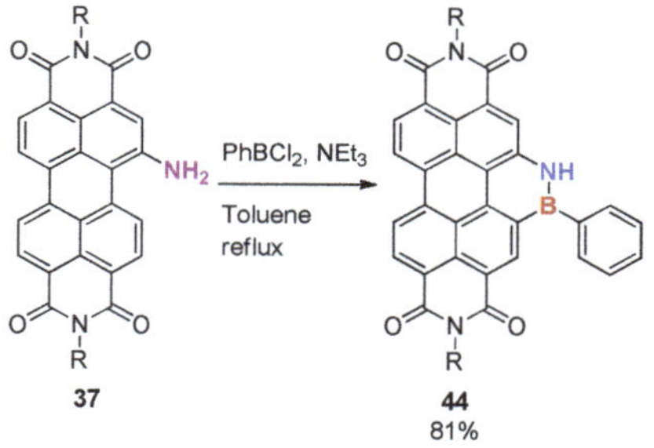

Scheme 13. B-N annulation between 1-aminoPDI **37** and dichlorophenylborane. R = cyclohexyl.

A combination of the Pictet–Spengler reaction and subsequent oxidative re-aromatization was reported as a way to access azabenzannulated PDIs **45** and **46** and N-decorated coronene bisimides derivatives (Scheme 14). Phenyl (42%) and 2-pyridyl (37%) groups were introduced by Z. Wang and colleagues by reacting aminoPDI **37** with the corresponding aldehyde in the presence of triflic acid (TfOH) in DMF at 110 °C [69]. This transformation was also performed by the group of F. Würthner using trifluoroacetic acid (TFA) as the catalyst [72]. In this reaction, an iminium undergoes an intramolecular electrophilic aromatic substitution and subsequent oxidative re-aromatization under an oxygen atmosphere. These compounds are characterized by low LUMO levels, making them good electron acceptors, and their absorption maximum being blue-shifted compared to regular PDI. Interestingly, PDI-pyridyl based iridium and ruthenium complexes were prepared and characterized; in particular, their phosphorescence in the near-infrared region [73,74].

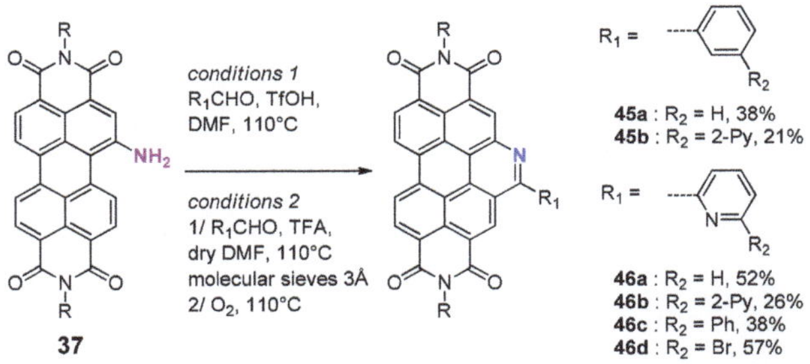

Scheme 14. Preparation of azabenzannulated PDI via a Pictet-Spengler approach. R = 2,6-diisopropylphenyl.

Recently, our group reported an alternative to the Pictet–Spengler reaction to access azabenzannulated PDIs **47** (Scheme 15) [35]. This original reaction involves the formation of an imine intermediate, its photocyclization by visible light followed by re-aromatization using DDQ. These three steps can be performed in one pot process affording azabenzannulated PDI dyes, including dimers, in higher yields than those reported before.

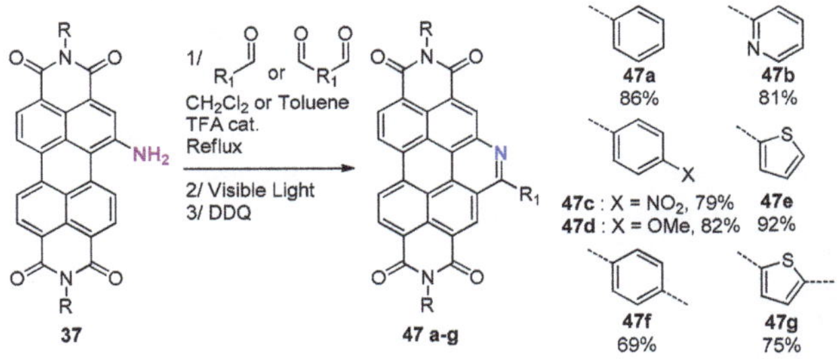

Scheme 15. Preparation of azabenzannulated PDI via an imine photocyclization. R = 1-hexylheptyl.

5. Conclusions

In this review, we have highlighted the synthetic use of 1-nitroPDI to modify the bay region of the perylenediimide (PDI) backbone and engineer unlimited photoredox modifications for various applications. Until now, PDI decorated with bromine atom in position 1 occupied a central role in the molecular design of a core-substituted-PDI based light-harvester and electron acceptor because of the versatility of the reactions involving aromatic halides. Nevertheless, as discussed in the manuscript, many reaction types have been now successfully applied to nitroPDI, from nucleophilic substitution, which could lead to an extension through annulation of the perylene core, to palladium catalyzed cross-couplings (Suzuki–Miyaura). This makes the derivative as attractive as bromoPDI. Besides, we have clearly demonstrated here the easier access to 1-nitroPDI (short reaction time, quantitative yield, green and fast purification) due to the higher selectivity of the mononitration compared to monobromination. Considering the two transformation steps, the synthetic methodology involving the nitration of PDI appears more appealing and should be more considered by the community, especially in regard to the possible industrialization.

Moreover, the classical reduction of the nitro group into amino can be easily applied. This group has not been widely exploited so far and the recent efficient imine formation and its subsequent photocyclization provides useful access to new azabenzannulated PDI based materials, promising candidates as non-fullerene acceptors (NFAs) in organic photovoltaics (OPV).

Author Contributions: All authors discussed, commented on and wrote the manuscript. All authors have read and agreed to the published version of the manuscript.

Funding: This research received no external funding".

Acknowledgments: Master's and PhD students who have actively participated to research on nitroPDI chemistry recently in our group are gratefully acknowledged.

Conflicts of Interest: The authors declare no conflict of interest.

References

1. Kardos, M. German Patent DE 276357, 1913.
2. Langhals, H. Cyclic carboxylic imide structures as structure elements of high stability. Novel developments in perylene dye chemistry. *Heterocycles* **1995**, *40*, 477–500. [CrossRef]
3. Zollinger, H. *Color Chemistry*, 3rd ed.; Wiley VCH: Weinheim, Germany, 2003.
4. Herbst, W.; Hunger, K. *Industrial Organic Pigments: Production, Properties, Applications*, 3rd ed.; Wiley VWH: Weinheim, Germany, 2004.
5. Loutfy, R.; Hor, A.; Kazmaier, P.; Tam, M. Layered organic photoconductive (OPC) devices incorporating N, N'-disubstituted diimide and bisarylimidazole derivatives of perylene-3,4,9,10-tetracarboxylic acid. *J. Imaging Sci.* **1989**, *33*, 151–159.
6. Soh, N.; Ueda, T. Perylene bisimide as a versatile fluorescent tool for environmental and biological analysis: A review. *Talanta* **2011**, *85*, 1233–1237. [CrossRef] [PubMed]
7. Quante, H.; Geerts, Y.; Müllen, K. Synthesis of Soluble Perylenebisamidine Derivatives. Novel Long-Wavelength Absorbing and Fluorescent Dyes. *Chem. Mater.* **1997**, *9*, 495–500. [CrossRef]
8. Sánchez, R.S.; Gras-Charles, R.; Bourdelande, J.L.; Guirado, G.; Hernando, J. Light- and Redox-Controlled Fluorescent Switch Based on a Perylenediimide–Dithienylethene Dyad. *J. Phys. Chem. C* **2012**, *116*, 7164–7172.
9. Yu, Z.; Wu, Y.; Liao, Q.; Zhang, H.; Bai, S.; Li, H.; Xu, Z.; Sun, C.; Wang, X.; Yao, J.; et al. Self-Assembled Microdisk Lasers of Perylenediimides. *J. Am. Chem. Soc.* **2015**, *137*, 15105–15111. [CrossRef]
10. Anthony, J.E. Small-Molecule, Nonfullerene Acceptors for Polymer Bulk Heterojunction Organic Photovoltaics. *Chem. Mater.* **2011**, *23*, 583–590. [CrossRef]
11. Li, C.; Wonneberger, H. Perylene Imides for Organic Photovoltaics: Yesterday, Today, and Tomorrow. *Adv. Mat.* **2012**, *24*, 613–636. [CrossRef]
12. Kozma, E.; Catellani, M. Perylene diimides based materials for organic solar cells. *Dyes Pigm.* **2013**, *98*, 160–179. [CrossRef]
13. Fernández-Lázaro, F.; Zink-Lorre, N.; Sastre-Santos, Á. Perylenediimides as non-fullerene acceptors in bulk-heterojunction solar cells (BHJSCs). *J. Mater. Chem. A* **2016**, *4*, 9336–9346. [CrossRef]
14. Wasielewski, M.R. Self-Assembly Strategies for Integrating Light Harvesting and Charge Separation in Artificial Photosynthetic Systems. *Acc. Chem. Res.* **2009**, *42*, 1910–1921. [CrossRef] [PubMed]
15. O'Neil, M.P.; Niemczyk, M.P.; Svec, W.A.; Gosztola, D.; Gaines, G.L., III; Wasielewski, M.R. Picosecond optical switching based on biphotonic excitation of an electron donor-acceptor-donor molecule. *Science* **1992**, *257*, 63–65. [CrossRef]
16. Prathapan, S.; Yang, S.I.; Seth, J.; Miller, M.A.; Bocian, D.F.; Holten, D.; Lindsey, J.S. Synthesis and Excited-State Photodynamics of Perylene-Porphyrin Dyads. 1. Parallel Energy and Charge Transfer via a Diphenylethyne Linker. *J. Phys. Chem. B* **2001**, *105*, 8237–8248. [CrossRef]
17. Serin, J.M.; Brousmiche, D.W.; Fréchet, J.M.J. Cascade energy transfer in a conformationally mobile multichromophoric dendrimer. *Chem. Commun.* **2002**, 2605–2607. [CrossRef] [PubMed]
18. Fukuzumi, S.; Ohkubo, K.; Ortiz, J.; Gutierrez, A.M.; Fernandez-Lazaro, F.; Sastre-Santos, A. Formation of a long-lived charge-separated state of a zinc phthalocyanine-perylenediimide dyad by complexation with magnesium ion. *Chem. Commun.* **2005**, 3814–3816. [CrossRef]
19. An, Z.; Odom, S.A.; Kelley, R.F.; Huang, C.; Zhang, X.; Barlow, S.; Padilha, L.A.; Fu, J.; Webster, S.; Hagan, D.J.; et al. Synthesis and Photophysical Properties of Donor- and Acceptor-Substituted 1,7-Bis(arylalkynyl)perylene-3,4:9,10-bis(dicarboximide)s. *J. Phys. Chem. A* **2009**, *113*, 5585–5593. [CrossRef]
20. Shoaee, S.; An, Z.; Zhang, X.; Barlow, S.; Marder, S.R.; Duffy, W.; Heeney, M.; McCulloch, I.; Durrant, J.R. Charge photogeneration in polythiophene-perylene diimide blend films. *Chem. Commun.* **2009**, 5445–5447. [CrossRef]

21. Rocard, L.; Berezin, A.; De Leo, F.; Bonifazi, D. Templated Chromophore Assembly by Dynamic Covalent Bonds. *Angew. Chem. Int. Ed.* **2015**, *54*, 15739–15743. [CrossRef]
22. Rocard, L.; Wragg, D.; Jobbins, S.A.; Luciani, L.; Wouters, J.; Leoni, S.; Bonifazi, D. Templated Chromophore Assembly on Peptide Scaffolds: A Structural Evolution. *Chem. Eur. J.* **2018**, *24*, 16136–16143. [CrossRef]
23. Yukruk, F.; Dogan, A.L.; Canpinar, H.; Guc, D.; Akkaya, E.U. Water-Soluble Green Perylenediimide (PDI) Dyes as Potential Sensitizers for Photodynamic Therapy. *Org. Lett.* **2005**, *7*, 2885–2887. [CrossRef]
24. Sun, M.; Müllen, K.; Yin, M. Water-soluble perylenediimides: Design concepts and biological applications. *Chem. Soc. Rev.* **2016**, *45*, 1513–1528. [CrossRef] [PubMed]
25. Roncali, J. Synthetic principles for bandgap control in linear π-conjugated systems. *Chem. Rev.* **1997**, *97*, 173–205. [CrossRef] [PubMed]
26. Roncali, J. Molecular Engineering of the Band Gap of π-Conjugated Systems: Facing Technological Applications. *Macromol. Rapid Commun.* **2007**, *28*, 1761–1775. [CrossRef]
27. Miletić, T.; Fermi, A.; Orfanos, I.; Avramopoulos, A.; De Leo, F.; Demitri, N.; Bergamini, G.; Ceroni, P.; Papadopoulos, M.G.; Couris, S.; et al. Tailoring Colors by O Annulation of Polycyclic Aromatic Hydrocarbons. *Chem. Eur. J.* **2017**, *23*, 2363–2378. [CrossRef]
28. Huang, C.; Barlow, S.; Marder, S.R. Perylene-3,4,9,10-tetracarboxylic Acid Diimides: Synthesis, Physical Properties, and Use in Organic Electronics. *J. Org. Chem.* **2011**, *76*, 2386–2407. [CrossRef]
29. Nowak-Krol, A.; Würthner, F. Progress in the synthesis of perylene bisimide dyes. *Org. Chem. Front.* **2019**, *6*, 1272–1318. [CrossRef]
30. Chen, K.-Y.; Chang, C.-W. Highly soluble monoamino-substituted perylene tetracarboxylic dianhydrides: Synthesis, optical and electrochemical properties. *Int. J. Mol. Sci.* **2014**, *15*, 22642–22660. [CrossRef]
31. Tsai, H.-Y.; Chang, C.-W.; Chen, K.-Y. 1,6- and 1,7-regioisomers of dinitro- and diamino-substituted perylene bisimides: Synthesis, photophysical and electrochemical properties. *Tetrahedron Lett.* **2014**, *55*, 884–888. [CrossRef]
32. Wang, R.; Li, G.; Zhang, A.; Wang, W.; Cui, G.; Zhao, J.; Shi, Z.; Tang, B. Efficient energy-level modification of novel pyran-annulated perylene diimides for photocatalytic water splitting *Chem. Commun.* **2017**, *53*, 6918–6921. [CrossRef]
33. Rocard, L.; Hatych, D.; Chartier, T.; Cauchy, T.; Hudhomme, P. Original Suzuki–Miyaura Coupling Using Nitro Derivatives for the Synthesis of Perylenediimide-Based Multimers. *Eur. J. Org. Chem.* **2019**, *2019*, 7635–7643. [CrossRef]
34. Gupta, R.K.; Shankar Rao, D.S.; Prasad, S.K.; Achalkumar, A.S. Columnar Self-Assembly of Electron-Deficient Dendronized Bay-Annulated Perylene Bisimides. *Chem. Eur. J.* **2018**, *24*, 3566–3575. [CrossRef] [PubMed]
35. Goujon, A.; Rocard, L.; Cauchy, T.; Hudhomme, P. An Efficient Imine Photocyclization as an Alternative to the Pictet-Spengler Reaction for the Synthesis of AzaBenzannulated Perylenediimide Dyes. *ChemRxiv* **2020**. [CrossRef]
36. Würthner, F. Bay-substituted perylene bisimides: Twisted fluorophores for supramolecular chemistry. *Pure Appl. Chem.* **2006**, *78*, 2341–2349. [CrossRef]
37. Rajasingh, P.; Cohen, R.; Shirman, E.; Shimon, L.J.W.; Rybtchinski, B. Selective Bromination of Perylene Diimides under Mild Conditions. *J. Org. Chem.* **2007**, *72*, 5973–5979. [CrossRef]
38. Langhals, H.; Kirner, S. Novel fluorescent dyes by the extension of the core of perylenetetracarboxylic bisimides. *Eur. J. Org. Chem.* **2000**, 365–380. [CrossRef]
39. Dominguez, C.; Baena, M.J.; Coco, S.; Espinet, P. Perylenecarboxydiimide-gold(I) organometallic dyes. Optical properties and Langmuir films. *Dyes Pigm.* **2017**, *140*, 375–383. [CrossRef]
40. Chen, K.-Y.; Chow, T.J. 1,7-Dinitroperylene bisimides: Facile synthesis and characterization as n-type organic semiconductors. *Tetrahedron Lett.* **2010**, *51*, 5959–5963. [CrossRef]
41. Chen, K.-Y.; Fang, T.-C.; Chang, M.-J. Synthesis, photophysical and electrochemical properties of 1-aminoperylene bisimides. *Dyes Pigm.* **2012**, *92*, 517–523. [CrossRef]
42. Chen, Z.J.; Wang, L.M.; Zou, G.; Zhang, L.; Zhang, G.J.; Cai, X.F.; Teng, M.S. Colorimetric and ratiometric fluorescent chemosensor for fluoride ion based on perylene diimide derivatives. *Dyes Pigm.* **2012**, *94*, 410–415. [CrossRef]
43. Wang, Y.; Zhang, L.; Zhang, G.; Wu, Y.; Wu, S.; Yu, J.; Wang, L. A new colorimetric and fluorescent bifunctional probe for Cu2+ and F- ions based on perylene bisimide derivatives. *Tetrahedron Lett.* **2014**, *55*, 3218–3222. [CrossRef]

44. Tsai, H.-Y.; Chen, K.-Y. Synthesis and optical properties of novel asymmetric perylene bisimides. *J. Lumin.* **2014**, *149*, 103–111. [CrossRef]
45. Rosenne, S.; Grinvald, E.; Shirman, E.; Neeman, L.; Dutta, S.; Bar-Elli, O.; Ben-Zvi, R.; Oksenberg, E.; Milko, P.; Kalchenko, V.; et al. Self-Assembled Organic Nanocrystals with Strong Nonlinear Optical Response. *Nano Lett.* **2015**, *15*, 7232–7237. [CrossRef] [PubMed]
46. Meng, D.; Sun, D.; Zhong, C.; Liu, T.; Fan, B.; Huo, L.; Li, Y.; Jiang, W.; Choi, H.; Kim, T.; et al. High-Performance Solution-Processed Non-Fullerene Organic Solar Cells Based on Selenophene-Containing Perylene Bisimide Acceptor. *J. Am. Chem. Soc.* **2016**, *138*, 375–380. [CrossRef] [PubMed]
47. Singh, P.; Kumar, K.; Bhargava, G.; Kumar, S. Self-assembled nanorods of bay functionalized perylenediimide: Cu^{2+} based 'turn-on' response for INH, complementary NOR/OR and TRANSFER logic functions and fluorosolvatochromism. *J. Mater. Chem. C* **2016**, *4*, 2488–2497. [CrossRef]
48. Kumar, K.; Bhargava, G.; Kumar, S.; Singh, P. Dissymmetric Bay-Functionalized Perylenediimides. *Synlett* **2018**, *29*, 1693–1699.
49. Singh, P.; Mittal, L.S.; Vanita, V.; Kumar, K.; Walia, A.; Bhargava, G.; Kumar, S. Self-assembled vesicle and rod-like aggregates of functionalized perylene diimide: Reaction-based near-IR intracellular fluorescent probe for selective detection of palladium. *J. Mater. Chem. B* **2016**, *4*, 3750–3759. [CrossRef]
50. Hendsbee, A.D.; Sun, J.-P.; Law, W.K.; Yan, H.; Hill, I.G.; Spasyuk, D.M.; Welch, G.C. Synthesis, Self-Assembly, and Solar Cell Performance of N-Annulated Perylene Diimide Non-Fullerene Acceptors. *Chem. Mater.* **2016**, *28*, 7098–7109. [CrossRef]
51. El-Berjawi, R.; Hudhomme, P. Synthesis of a perylenediimide-fullerene C60 dyad: A simple use of a nitro leaving group for a Suzuki-Miyaura coupling reaction. *Dyes Pigm.* **2018**, *159*, 551–556. [CrossRef]
52. Hruzd, M.; Rocard, L.; Allain, M.; Hudhomme, P. Suzuki-Miyaura Coupling on Dinitro-bay Substituted Perylenediimide. 2020; Unpublished results; manuscript in preparation.
53. Tsai, H.-Y.; Chang, C.-W.; Chen, K.-Y. 1,6- and 1,7-regioisomers of asymmetric and symmetric perylene bisimides: Synthesis, characterization and optical properties. *Molecules* **2014**, *19*, 327–341. [CrossRef]
54. Dubey, R.K.; Efimov, A.; Lemmetyinen, H. 1,7- And 1,6-Regioisomers of Diphenoxy and Dipyrrolidinyl Substituted Perylene Diimides: Synthesis, Separation, Characterization, and Comparison of Electrochemical and Optical Properties. *Chem. Mater.* **2011**, *23*, 778–788. [CrossRef]
55. Kong, X.; Gao, J.; Ma, T.; Wang, M.; Zhang, A.; Shi, Z.; Wei, Y. Facile synthesis and replacement reactions of mono-substituted perylene bisimide dyes. *Dyes Pigm.* **2012**, *95*, 450–454. [CrossRef]
56. Sun, D.; Meng, D.; Cai, Y.; Fan, B.; Li, Y.; Jiang, W.; Huo, L.; Sun, Y.; Wang, Z. Non-Fullerene-Acceptor-Based Bulk-Heterojunction Organic Solar Cells with Efficiency over 7%. *J. Am. Chem. Soc.* **2015**, *137*, 11156–11162. [CrossRef] [PubMed]
57. Nazari, M.; Cieplechowicz, E.; Welsh, T.A.; Welch, G.C. A direct comparison of monomeric vs. dimeric and non-annulated vs. N-annulated perylene diimide electron acceptors for organic photovoltaics. *New J. Chem.* **2019**, *43*, 5187–5195. [CrossRef]
58. Ma, Z.; Fu, H.; Meng, D.; Jiang, W.; Sun, Y.; Wang, Z. Isomeric N-Annulated Perylene Diimide Dimers for Organic Solar Cells. *Chem. Asian J.* **2018**, *13*, 918–923. [CrossRef] [PubMed]
59. Cann, J.R.; Cabanetos, C.; Welch, G.C. Synthesis of Molecular Dyads and Triads Based Upon N-Annulated Perylene Diimide Monomers and Dimers. *Eur. J. Org. Chem.* **2018**, *2018*, 6933–6943. [CrossRef]
60. Chen, L.; Xia, P.; Du, T.; Deng, Y.; Xiao, Y. Catalyst-Free One-Pot Synthesis of Unsymmetrical Five- and Six-Membered Sulfur-Annulated Heterocyclic Perylene Diimides for Electron-Transporting Property. *Org. Lett.* **2019**, *21*, 5529–5532. [CrossRef]
61. Ma, Y.; Shi, Z.; Zhang, A.; Li, J.; Wei, X.; Jiang, T.; Li, Y.; Wang, X. Self-assembly, optical and electrical properties of five membered O- or S-heterocyclic annulated perylene diimides. *Dyes Pigm.* **2016**, *135*, 41–48. [CrossRef]
62. Wang, R.; Li, G.; Zhou, Y.; Hao, P.; Shang, Q.; Wang, S.; Zhang, Y.; Li, D.; Yang, S.; Zhang, Q. Facile Syntheses, Characterization, and Physical Properties of Sulfur-Decorated Pyran-Annulated Perylene Diimides. *Asian J. Org. Chem.* **2018**, *7*, 702–706. [CrossRef]
63. Li, X.; Wang, H.; Nakayama, H.; Wei, Z.; Schneider, J.A.; Clark, K.; Lai, W.-Y.; Huang, W.; Labram, J.G.; de Alaniz, J.R.; et al. Multi-Sulfur-Annulated Fused Perylene Diimides for Organic Solar Cells with Low Open-Circuit Voltage Loss. *ACS Appl. Energy Mater.* **2019**, *2*, 3805–3814. [CrossRef]

64. Yang, Y. Palladium-Catalyzed Cross-Coupling of Nitroarenes. *Angew. Chem. Int. Ed.* **2017**, *56*, 15802–15804. [CrossRef]
65. Yadav, M.R.; Nagaoka, M.; Kashihara, M.; Zhong, R.-L.; Miyazaki, T.; Sakaki, S.; Nakao, Y. The Suzuki-Miyaura Coupling of Nitroarenes. *J. Am. Chem. Soc.* **2017**, *139*, 9423–9426. [CrossRef] [PubMed]
66. Rocard, L.; Hudhomme, P. Recent developments in the Suzuki-Miyaura reaction using nitroarenes as electrophilic coupling reagents. *Catalysts* **2019**, *9*, 213. [CrossRef]
67. Aivali, S.; Tsimpouki, L.; Anastasopoulos, C.; Kallitsis, J.K. Synthesis and optoelectronic characterization of perylene diimide-quinoline based small molecules. *Molecules* **2019**, *24*, 4406. [CrossRef] [PubMed]
68. Tsai, H.-Y.; Chen, K.-Y. 1,7-Diaminoperylene bisimides: Synthesis, optical and electrochemical properties. *Dyes Pigm.* **2013**, *96*, 319–327. [CrossRef]
69. Hao, L.; Jiang, W.; Wang, Z. Integration of nitrogen into coronene bisimides. *Tetrahedron* **2012**, *68*, 9234–9239. [CrossRef]
70. Li, G.; Zhao, Y.; Li, J.; Cao, J.; Zhu, J.; Sun, X.W.; Zhang, Q. Synthesis, Characterization, Physical Properties, and OLED Application of Single BN-Fused Perylene Diimide. *J. Org. Chem.* **2015**, *80*, 196–203. [CrossRef]
71. Zhang, L.; Wang, Y.; Yu, J.; Zhang, G.; Cai, X.; Wu, Y.; Wang, L. A colorimetric and fluorescent sensor based on PBIs for palladium detection. *Tetrahedron Lett.* **2013**, *54*, 4019–4022. [CrossRef]
72. Schulze, M.; Philipp, M.; Waigel, W.; Schmidt, D.; Wuerthner, F. Library of Azabenz-Annulated Core-Extended Perylene Derivatives with Diverse Substitution Patterns and Tunable Electronic and Optical Properties. *J. Org. Chem.* **2016**, *81*, 8394–8405. [CrossRef]
73. Shi, J.; Fan, J.; Qu, Z.; Wang, S.; Wang, Y. Solution concentration-dependent tunable emission in cyclometalated iridium complex bearing perylene diimide (PDI) ligand: From visible to near-infrared emission. *Dyes Pigm.* **2018**, *154*, 263–268. [CrossRef]
74. Schulze, M.; Steffen, A.; Wuerthner, F. Near-IR Phosphorescent Ruthenium(II) and Iridium(III) Perylene Bisimide Metal Complexes. *Angew. Chem. Int. Ed.* **2015**, *54*, 1570–1573. [CrossRef]

© 2020 by the authors. Licensee MDPI, Basel, Switzerland. This article is an open access article distributed under the terms and conditions of the Creative Commons Attribution (CC BY) license (http://creativecommons.org/licenses/by/4.0/).

Review

Recent Progress in Nitro-Promoted Direct Functionalization of Pyridones and Quinolones

Feiyue Hao [1] and Nagatoshi Nishiwaki [2],*

[1] School of Pharmaceutical and Materials Engineering, Taizhou University, Jiaojiang 318000, China; haofeiyue@tzc.edu.cn

[2] School of Environmental Science and Engineering, Kochi University of Technology, Tosayamada, Kami, Kochi 782-8502, Japan

* Correspondence: nishiwaki.nagatoshi@kochi-tech.ac.jp; Tel.: +81-887-57-2517; Fax: +81-887-57-2520

Academic Editors: Nagatoshi Nishiwaki and Mathieu Pucheault
Received: 18 January 2020; Accepted: 3 February 2020; Published: 5 February 2020

Abstract: Nitro group is one of the most important functional groups in organic syntheses because its strongly electron-withdrawing ability activates the scaffold, facilitating the reaction with nucleophilic reagents or the Diels–Alder reaction. In this review, recent progress in the nitro-promoted direct functionalization of pyridones and quinolones is highlighted to complement previous reviews.

Keywords: nitro; pyridone; 1-methyl-2-quinolone; cycloaddition; direct functionalization

1. Introduction

Natural and synthetic aza-heterocycles represent an important class of organic compounds [1–5]. Among the large number of aza-heterocycles available, pyridones and quinolones, both of which have a common six-membered aza-framework, exhibit a wide range of pharmacologically important activities (Figure 1) [6–10]. Therefore, various methods for the preparation of structurally diverse pyridones and quinolones have been studied in detail [6,11–21].

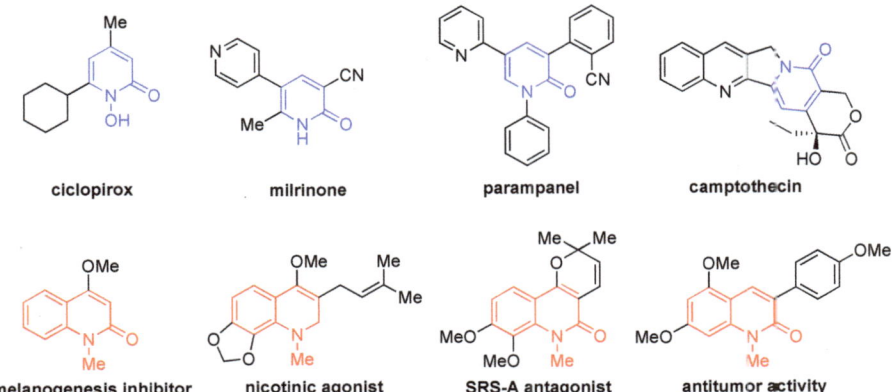

Figure 1. Biological activities of pyridones and quinolones.

Conventional strategies for the synthesis of aza-heterocycles involve (1) construction of aza-heterocycle frameworks from prefunctionalized starting materials, (2) ring transformation leading to aza-heterocycle frameworks, and (3) direct functionalization of aza-heterocycle frameworks, which are supplementary to each other (Figure 2) [22].

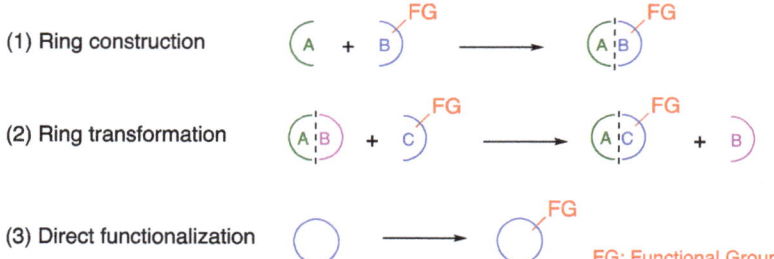

Figure 2. Conventional strategies for the functionalization of aza-heterocycles.

Among these three protocols, direct functionalization of aza-heterocycles, for preparing new diversely functionalized heterocycles, is the most efficient approach from a practical viewpoint, because it requires only simple experimental manipulations. Accordingly, the development of easy and efficient methods for the direct functionalization of quinolone and pyridone frameworks is highly demanded. However, only a few such methods are currently available because these scaffolds are inert due to the aromaticity (Figure 3) [22].

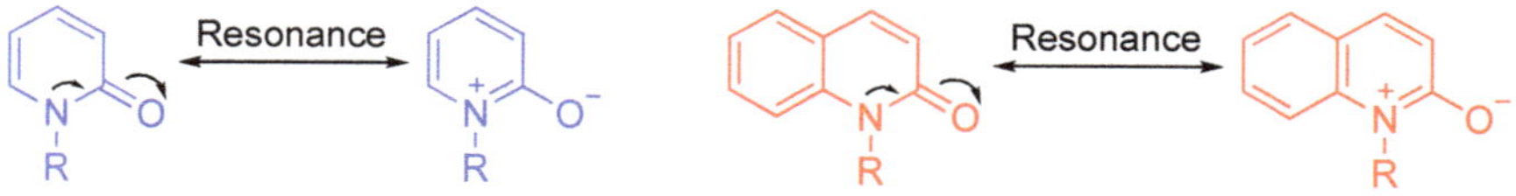

Figure 3. Resonance structure of pyridone framework.

To the best of our knowledge, the currently used methods for direct functionalization of the quinolone and pyridone scaffolds are mainly focused on transition-metal-catalyzed cross-coupling and C–H activation reactions [6,11–21]. However, most of these methods suffer from some limitations, such as the use of potentially poisonous and expensive noble metals, along with harsh reaction conditions.

However, the nitro group, which is often described as a "synthetic chameleon [23]," serves as a precursor for versatile functionalities, such as formyl, acyl, cyano, and amino groups (Scheme 1) [24–28]. Moreover, the nitro group has been proved to activate many different scaffolds because of its strong electron-withdrawing ability, facilitating the reaction with nucleophilic reagents [29,30]. The nitro group is also a good leaving group, which is often involved in addition–elimination reactions [31,32].

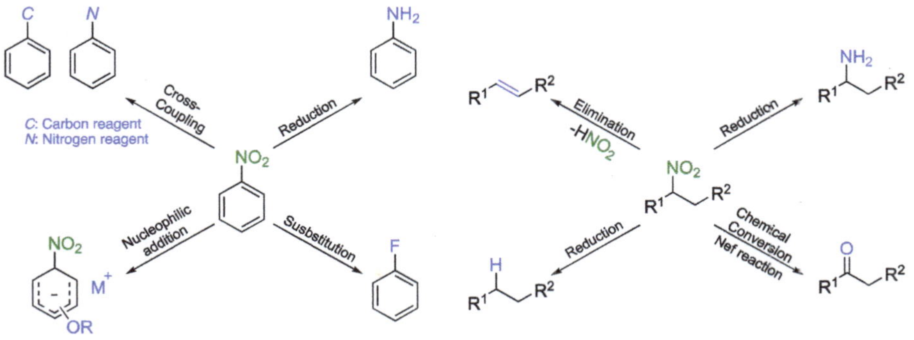

Scheme 1. Properties of a nitro group.

Based on these significant properties of the nitro group, the synthetic utility of nitrated aza-heterocycles in the preparation of functionalized aza-heterocycles has been widely investigated [33]. However, electrophilic nitration of pyridines and quinolines is difficult because of the electron deficiency of the aromatic cores. On the contrary, it is possible to nitrate pyridones and quinolones because the dearomatization of these scaffolds is easier than that of pyridines and quinolines. Indeed, the introduced nitro groups activate the scaffolds to facilitate direct functionalization, which affords structurally diverse aza-heterocycles. Herein, recent progress in the nitro-promoted direct functionalization of pyridones and quinolones in the past couple decades is highlighted.

2. Cycloaddition of Nitropyridones

The nitro group is a strongly electron-withdrawing group that reduces the electron density on the scaffold. Further, 2-pyridones possessing a nitro group are highly electron-deficient, and they serve as dienophiles that undergo Diels–Alder (D–A) cycloaddition with electron-rich dienes, forming fused aza-heterocycles [34].

When 5-nitro-2-pyridones **1** are reacted with 2,3-dimethyl-1,3-butadiene **2**, quinolones **3** are formed via regioselective D–A cycloaddition at the 5- and 6-positions and subsequent aromatization accompanied by elimination of nitrous acid (Table 1). For 5-nitropyridone bearing a methoxycarbonyl group at the 3-position, the D–A reaction occurs chemoselectively to yield the corresponding 3-functionalized quinolone **3c**.

Table 1. D-A cycloaddition of 5-nitropyridones **1** with diene **2**.

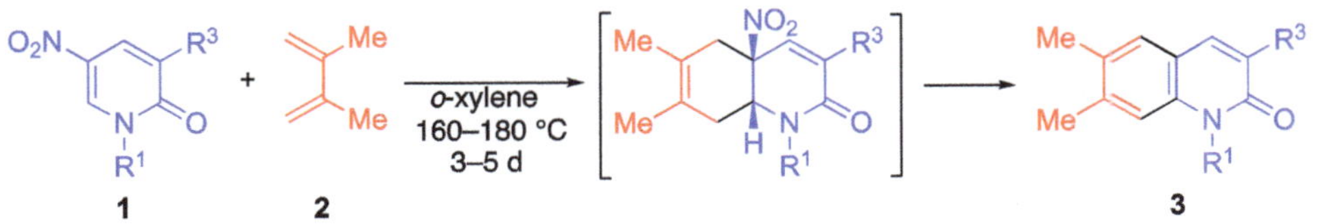

R^1	R^3		Yield/%
H	H	a	26
Me	H	b	30
Me	COOMe	c	22

It is known that 5-nitropyridones **4** possessing electron-withdrawing groups at the 3- and/or 4-positions have two electron-deficient sites on the ring. When these substrates are subjected to D–A reactions with diene **2**, the reaction proceeds stereoselectively to produce the functionalized *cis*-adducts

5 and **6**, accompanied by denitration (Table 2). Since the reaction is conducted under harsh conditions, the denitration of either pyridone **4b** or the cycloadducts **5'** and **6'** might occur (Scheme 2), however a detailed explanation has not been reported in the literature [34].

Table 2. D-A reaction of 5-nitropyridones **4** accompanying elimination of the nitro group.

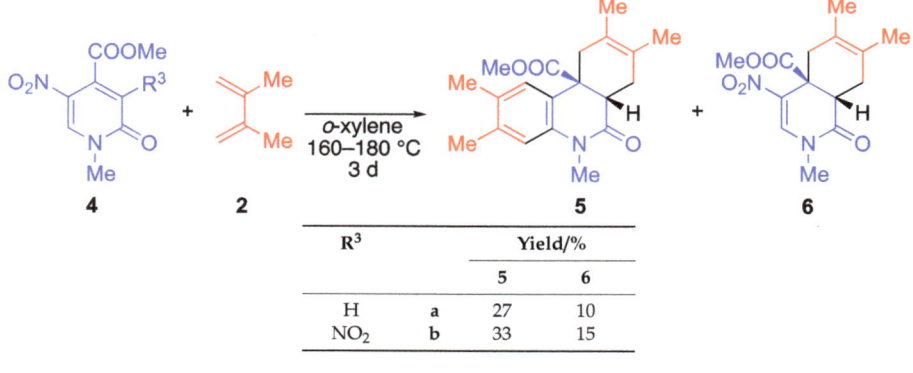

R³		Yield/%	
		5	6
H	a	27	10
NO₂	b	33	15

Scheme 2. Two plausible pathways for cycloadducts **5** and **6** including denitration.

D–A cycloaddition of 1-unsubstituted 3-nitro-2-pyridones **7a** with diene **2** yields the *cis*-condensed tetrahydroisoquinolone **8a** stereoselectively. For 1-methyl-3-nitro-2-pyridone **7b**, *cis*-tetrahydroisoquinolone **8b** as well as aromatized isoquinolone **9b** is formed via dehydrogenation and release of a nitrous acid. The use of a substrate with 4-methoxycarbonyl substitution affords *cis*-tetrahydroisoquinolone **8c** as the sole product (Table 3).

Table 3. Cycloaddition of 3-nitropyridones **7** with diene **2**.

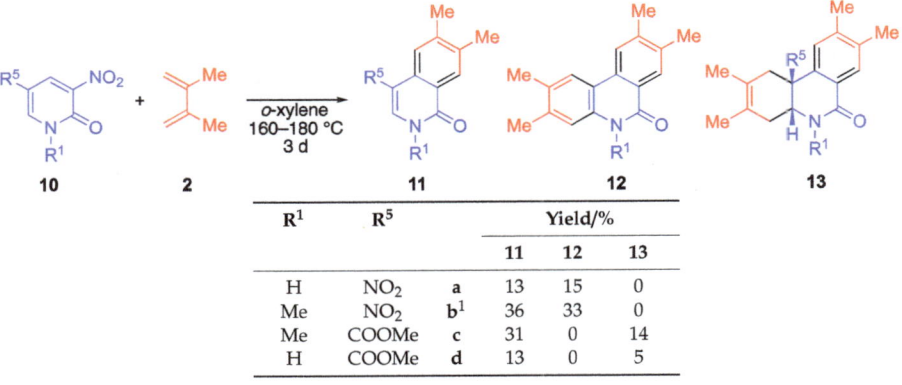

R¹	R⁴		Yield/%	
			8	9
H	H	a	0	15
Me	H	b	20	22
Me	COOMe	c	36	0

The reaction of 1-unsubstituted 3,5-dinitropyridone **10a** gives an aromatized isoquinolone **11a** via cycloaddition at the 3- and 4-positions, followed by dehydrogenation and elimination of nitrous acid; an aromatized phenanthridone **12a** is also obtained via double D–A adduct formation (Table 4). However, the reaction of 1-methyl-3,5-dinitro-2-pyridone **10b** furnishes not only 4-nitroisoquinolone **11b** and phenanthridone **12b**, but also *cis*-tetrahydroisoquinolone **8b**, via cycloaddition at the 3- and 4-positions accompanied by heating-promoted elimination of the nitro group at the 5-position. D–A reactions of 3-nitro-2-pyridones **10c** and **10d** with 5-methoxycarbonyl substitution mainly yield the aromatized isoquinolones **11c** and **11d**, respectively, in addition to the incompletely aromatized *cis*-phenanthridone adducts **13c** and **13d**, respectively.

Table 4. Cycloaddition of 5-substituted 3-nitropyridones **10** with diene **2**.

R¹	R⁵		Yield/%		
			11	12	13
H	NO₂	a	13	15	0
Me	NO₂	b¹	36	33	0
Me	COOMe	c	31	0	14
H	COOMe	d	13	0	5

¹ 8% of **8b** is obtained.

3. Cycloaddition of Nitroquinolones

The D–A reactions at the nitroalkene moiety of 3-nitrated 1-methyl-2-quinolones **14** with electron-rich dienes yield aromatized phenanthridone derivatives **15** (Table 5). Although this method enables simultaneous C–C bond formation at the 3- and 4-positions of the quinolone framework, harsh reaction conditions must be employed [35,36].

Table 5. Cycloaddition of 3-nitrated quinolones **14** with dienes **2**.

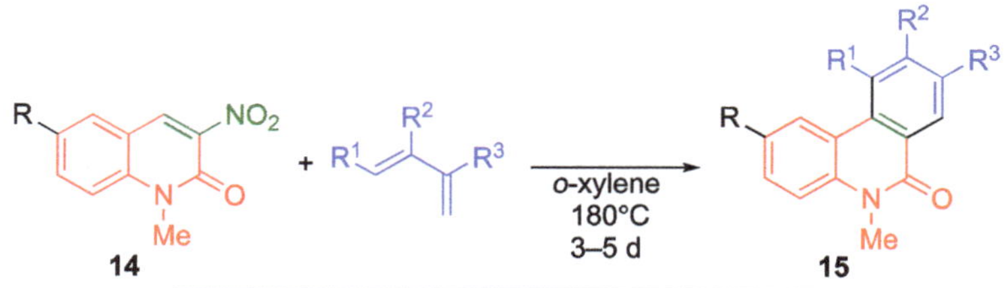

R	R¹	R²	R³		Yield/%
H	OMe	H	H	a	83
NO₂	OMe	H	H	b	68
H	H	Me	Me	c	95
NO₂	H	Me	Me	d	64
H	H	OMe	OMe	e	45
NO₂	H	OMe	OMe	f	13

On the contrary, 1-methyl-3,6,8-trinitro-2-quinolone **16** undergoes cycloaddition with dienes easily under mild conditions (Scheme 3). Indeed, the cycloaddition of **16** with cyclopentadiene proceeds smoothly to furnish a tetracyclic compound **17** that aromatizes via elimination of a nitrous acid in the presence of triethylamine to afford compound **18** [37]. Similarly, the cycloaddition using α,β-unsaturated oxime, instead of cyclopentadiene, as a heterodiene affords the polycyclic diazaphenanthrene **19** (Scheme 4) [38].

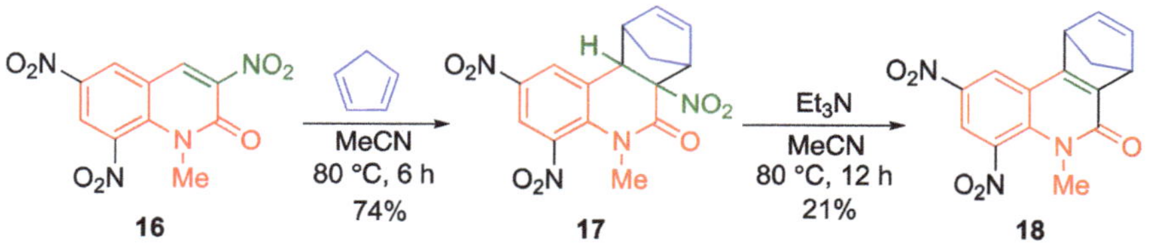

Scheme 3. Diels-Alder cycloaddition of trinitroquinolone **16** with cyclopentadiene.

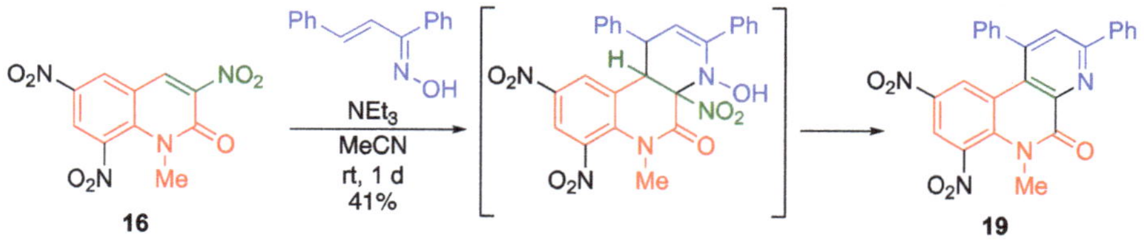

Scheme 4. Cycloaddition of **16** with α,β-unsaturated oxime.

The high reactivity of trinitroquinolone **16** is due to the steric repulsion between the 1-methyl and 8-nitro groups, disturbing the coplanarity of the pyridone moiety and the benzene ring. Consequently, the pyridone ring of **16** loses its aromaticity and serves as an activated nitroalkene (Figure 4) [39].

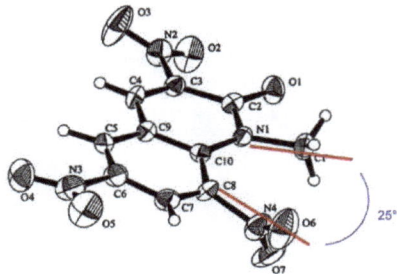

Figure 4. ORTEP (30% probability ellipsoids) view of trinitroquinolone **16**.

A nitroalkene shows dual behavior in cycloaddition reactions (Figure 5). In reaction with a diene, the nitroalkene serves as a dienophile to form a cyclohexene ring. On the other hand, it serves as a heterodiene in reaction with an electron-rich alkene to construct an oxazine ring. The nitroalkene moiety of trinitroquinolone **16** also serves as a heterodiene in the reaction with ethoxyethene to construct a fused oxazine ring **20** (Scheme 5) [38], which yields an acetal **21** via ring-opening reaction upon treatment with alcohol under reflux conditions.

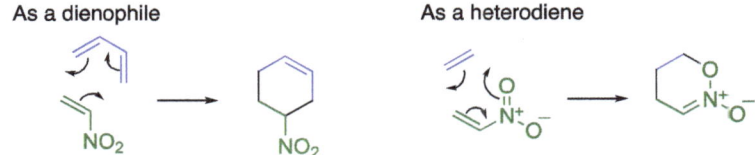

Figure 5. Dual behaviors of a nitroalkene in the cycloaddition reaction.

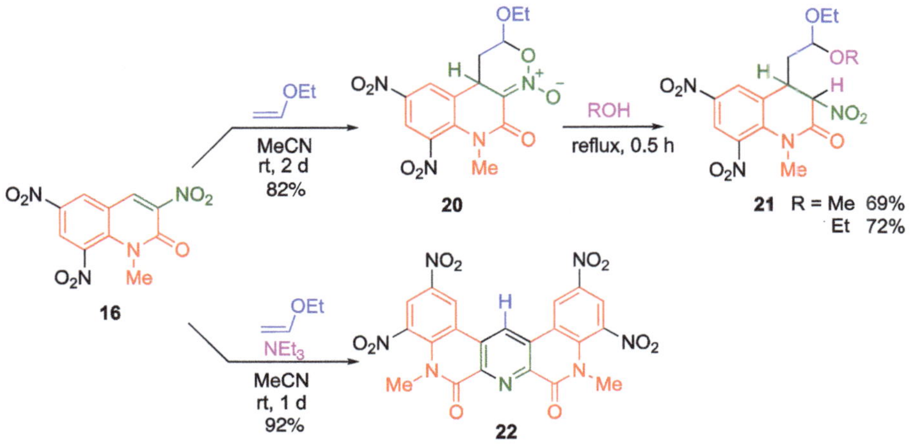

Scheme 5. Cycloaddition of **16** with ethoxyethane.

Interestingly, a quinolino[3,4-*b*][1,9]diazaphenanthrene derivative **22** is formed when the same reaction is conducted in the presence of triethylamine (Scheme 5) [38]. A plausible mechanism is shown in Scheme 6. After forming the cyclic nitronate **20**, triethylamine assists the proton transfer from the 4-position to the anionic oxygen of the nitronate. The subsequent retro D–A reaction gives the α,β-unsaturated oxime **A**, accompanied by a loss of ethyl formate. Oxime **A** serves as an electron-rich heterodiene that undergoes cycloaddition with another molecule of **16** to afford a new pyridine ring, and subsequent aromatization and elimination of nitrous acid and water furnishes the polycyclic product **22**. In this reaction, two molecules of trinitroquinolone **16** undergo two kinds of cycloaddition reactions: one molecule serves as a heterodiene and the other serves as a dienophile. This is the first example of a nitroalkene that exhibits dual behavior in the same reaction mixture (Figure 5).

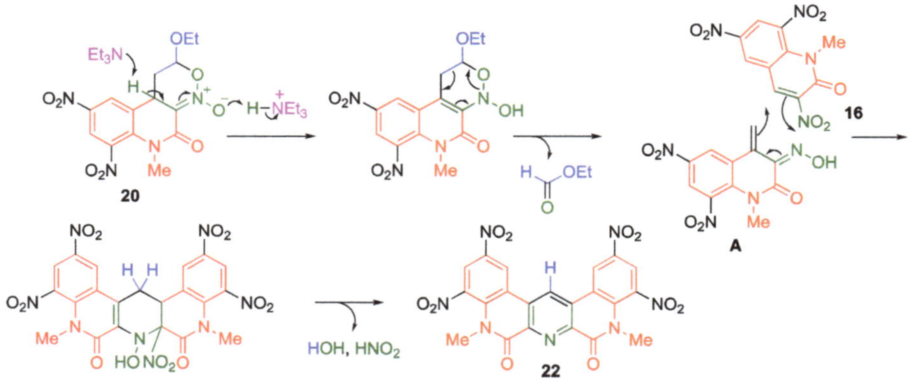

Scheme 6. A plausible mechanism for the formation of product **22**.

4. Nitro-Promoted Cyclization of Pyridones via Nucleophilic Addition

The strongly electron-withdrawing ability of the nitro group activates the scaffold for nucleophilic attack at the vicinal position on the nitroalkene. The nitroalkene moiety of nitropyridones is also susceptible to nucleophilic reaction. Indeed, 1-substituted nitropyridones **23** and **24** react with ethyl isocyanoacetate in the presence of 1,8-diazabicyclo[5.4.0]undec-7-ene (DBU) to afford the pyrrolopyridine derivatives **25** and **26**, respectively (Scheme 7) [40]. In the latter case, nucleophilic attack of isocyanoacetate occurs regioselectively at the 6-position.

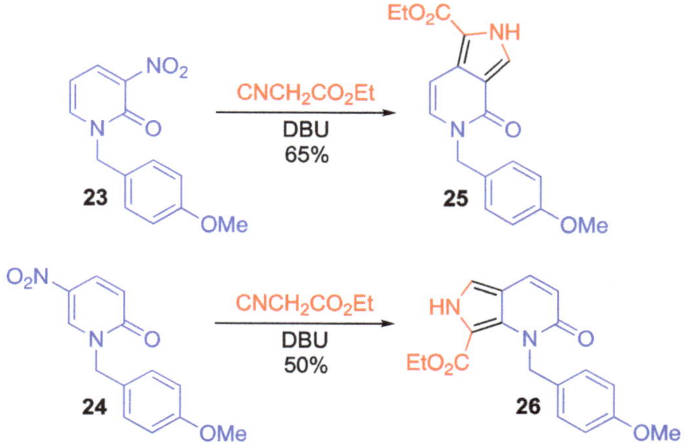

Scheme 7. Cyclization of nitropyridones **23** and **24**.

The reaction is initiated by the nucleophilic addition of isocyanoacetate to nitropyridone under basic conditions to produce an anionic intermediate stabilized by the nitro group (Scheme 8). Then, the nucleophilic attack of the nitronate to the protonated isocyano group affords dihydro-2H-pyrrole, from which a pyrrole ring is produced via aromatization by elimination of nitrous acid.

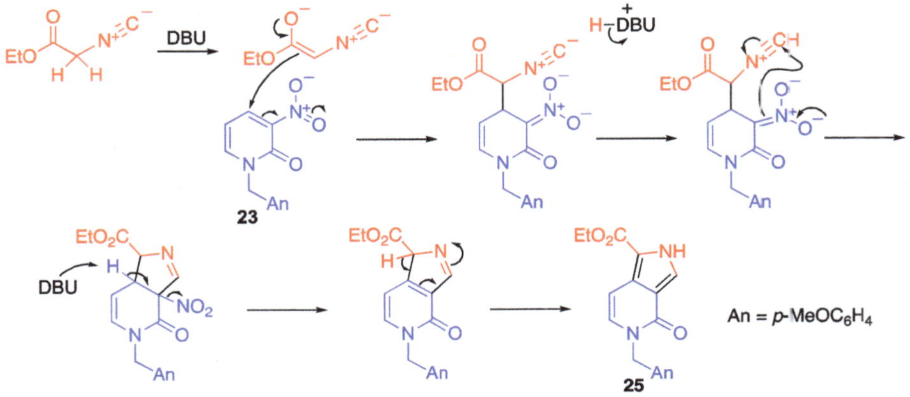

Scheme 8. A plausible mechanism for cyclization of nitropyridone 23 with isocyanoacetate.

5. Nitro-Promoted Direct Functionalization of Quinolones

5.1. Direct C–C Bond Formation at the 4-Position via Cine-Substitution

To the best of our knowledge, the currently used methods for direct C–C bond formation in 1-methyl-2-quinolone (**MeQone**) framework are mainly limited to transition-metal-catalyzed cross-coupling or C–H activation reactions [11–16]. As an alternative, the introduction of a nitro group has proved helpful in facilitating direct functionalization of the **MeQone** framework, affording diversely functionalized **MeQones**. Indeed, *cine*-substitution of trinitroquinolone **16** with various nucleophiles can easily proceed to afford 4-functionalized 6,8-dinitro-1-methyl-2-quinolones (**4FDNQ**) [22]. Initially, the nucleophilic substitution proceeds at the 4-position of **16** to form an adduct intermediate; then, a proton is transferred from the basic group to the 3-position of the adduct intermediate, affording 3,4-dihydroquinolone. The subsequent elimination of nitrous acid, accompanied by aromatization, yields **4FDNQ** (Scheme 9). This reaction enables regioselective functionalization at the 4-position of the **MeQone** framework. Direct C–C bond formation at the 4-position of the **MeQone** framework is easily achieved upon treatment of **16** with carbon nucleophiles, including 1,3-dicarbonyl compounds, nitroalkanes, aldehydes/ketones, enamines, cyanides, and phenoxides, leading to the formation of versatile skeletons.

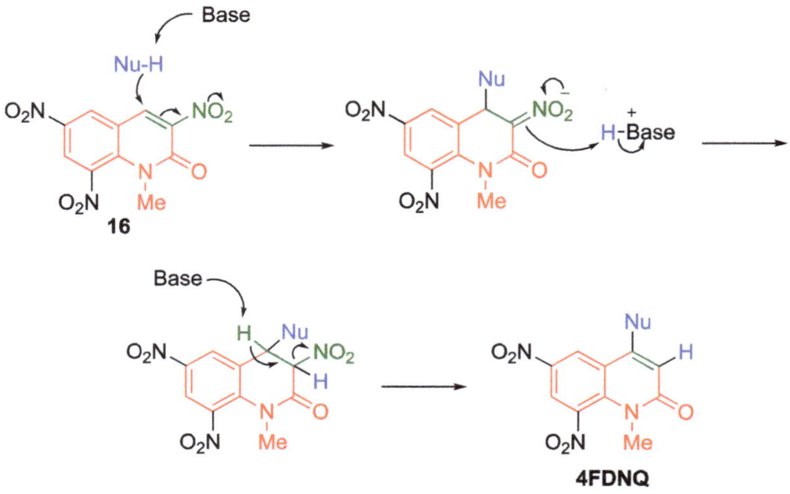

Scheme 9. *cine*-Substitution of trinitroquinolone **16**.

5.1.1. *cine*-Substitution of Trinitroquinolone with 1,3-dicarbonyl Compounds

When trinitroquinolone **16** is reacted with 1,3-dicarbonyl compounds in the presence of triethylamine, 4-position functionalization is efficiently achieved via *cine*-substitution (Table 6) [41]. Diketones, keto esters, and diesters can be used as nucleophiles in this reaction to afford the corresponding products **27a–e**.

Table 6. *cine*-Substitution of **16** with 1,3-dicarbonyl compounds.

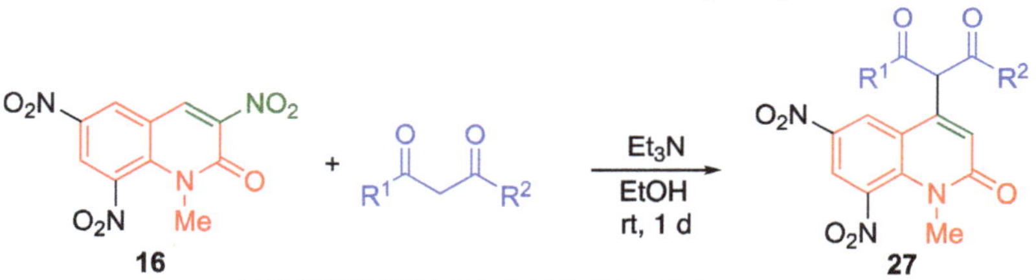

R^1	R^2		Yield/%
Me	Me	a	88
-(CH$_2$)$_3$-		b	68
Me	OEt	c	93
CH$_2$COOEt	OEt	d	26
OEt	OEt	e	93

When the nitro group at the 8-position is removed, no reaction occurs, even under heating. On the other hand, *cine*-substitution proceeds smoothly even upon replacement of the electron-withdrawing nitro group of **16** with an electron-donating methyl group (Table 7). These results indicate that the steric repulsion of this substituent with the 1-methyl group activates the **MeQone** framework, as mentioned in Section 3 [42].

Table 7. Effect of the substituent at the 8-position for the *cine*-substitution.

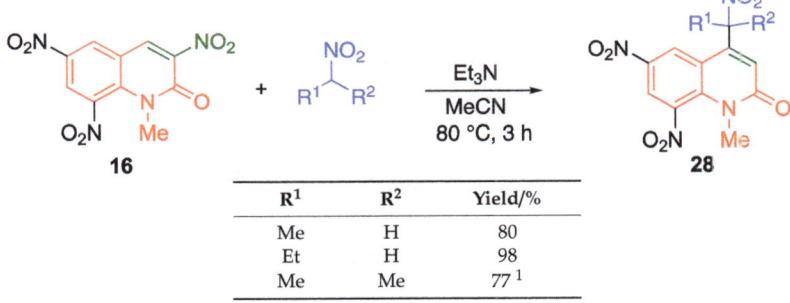

R	Yield/%
NO$_2$	88
H	0 [1]
Me	92

[1] At 80 °C.

5.1.2. *cine*-Substitution of Trinitroquinolone with Nitroalkanes

Nitroalkylation of trinitroquinolone **16** is also achieved by using a nitroalkane as a carbon nucleophile in the presence of triethylamine (Table 8) [43]. While primary nitroalkanes undergo *cine*-substitution efficiently at room temperature, secondary nitroalkanes with steric hindrance are less reactive, requiring longer reaction times and affording relatively low yields.

Table 8. *cine*-Substitution of **16** with nitroalkanes.

R^1	R^2	Yield/%
Me	H	80
Et	H	98
Me	Me	77 [1]

[1] For 1 d.

5.1.3. *cine*-Substitution of Trinitroquinolone with Aldehyde, Ketones and Enamines

Besides aldehydes, functionalized ketones, such as aliphatic, alicyclic, aromatic, and heteroaromatic ketones work well as carbon nucleophiles in the *cine*-substitution of trinitroquinolone **16**, giving acylmethylated products (Table 9) [44]. Since the acylmethyl group can serve as a scaffold for further chemical transformations, this method can be useful for the construction of a new library of compounds with **MeQone** framework.

Table 9. *cine*-Substitution of **16** with ketones.

R¹	R²	R³	Yield/%
H	Me	Me	41
Me	H	H	83
Ph	H	H	83
Et	Me	H	18
-(CH$_2$)$_4$-		H	82
Ph	Me	H	77
Ph	Ph	H	69
2-furyl	H	H	45
2-pyridyl	H	H	74

More-reactive enamines can also be used as nucleophiles instead of ketones, which undergo *cine*-substitution in the presence of water at room temperature. After the addition of enamine to trinitroquinolone **16**, hydrolysis of the formed iminium ion forms an acylmethyl group. In this case, the product is obtained as a morpholinium salt **30** (Table 10) [44].

Table 10. *cine*-Substitution of **16** with enamines.

R¹	R²	R³	Yield/%
H	Me	Me	98
-(CH$_2$)$_4$-		H	40
Ph	Me	H	43
Ph	Ph	H	98

5.1.4. *cine*-Substitution of Trinitroquinolone with Phenoxides

A combination of electrophilic trinitroquinolone **16** and nucleophilic phenoxide ions results in direct arylation of the **MeQone** framework (Figure 6) [45]. When **16** is treated with potassium phenoxides possessing electron-donating groups, double *cine*-substitution proceeds to afford bis(quinolyl)phenols **31** and **32**. On the other hand, sterically hindered or electron-deficient phenoxides give monoquinolylphenols **33** and **34** as the only products. Since direct introduction of an aryl group into the **MeQone** framework is difficult, this method is considered one of the more useful modifications.

From another viewpoint, trinitroquinolone is an aromatic compound. Hence, this reaction can be regarded as an electrophilic arylation, which is not achieved in the usual Friedel–Crafts reaction. This transformation is initiated by the nucleophilic addition of phenoxide at the 4-position of **16** (Scheme 10).

The newly introduced benzene ring is aromatized with the assistance of another phenoxide. In addition, proton transfer from the 4-position to an adjacent position of the quinolone ring occurs to afford the dianionic intermediate **B**. Since **B** is a highly electron-rich species, it immediately attacks another molecule of **16** to afford bis(quinolyl)phenols **31** (path **a**). On the other hand, protonation of **B** followed by elimination of nitrous acid is the preferred route to furnish monoquinolylphenol when electron-deficient or bulky phenoxides are used (path **b**).

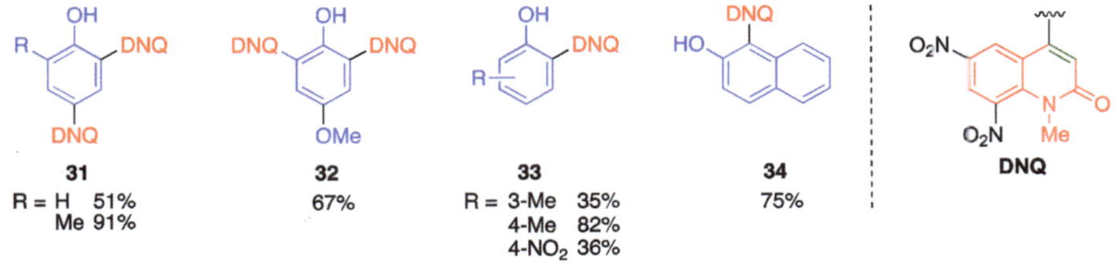

Figure 6. *cine*-Substituted products from **16** and potassium phenoxides.

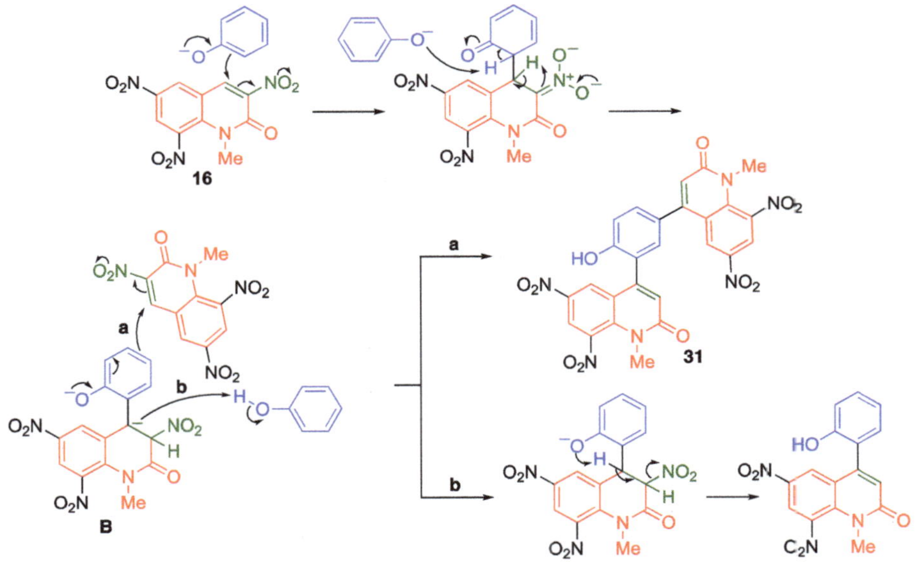

Scheme 10. A plausible mechanism for the reaction of **16** with phenoxide.

5.1.5. *cine*-Substitution of Trinitroquinolone with Cyanides

Nitriles represent an important structural motif in medicinal chemistry because of their versatile biological activities [46]. In addition, they have been recognized as extremely useful intermediates for the preparation of other useful building blocks [47–49]. Therefore, considerable research effort has been dedicated to the development of methods for introducing cyano groups into organic molecules. Inspired by the above methods for direct C–C bond formation on the **MeQone** framework, researchers have used potassium cyanide as a carbon nucleophile for reacting with trinitroquinolone **16** to prepare 4-cyano-2-quinolone derivative **35** (Scheme 11) [42]. In this reaction, dimeric product **36** is also obtained. After the addition of a cyanide to **16**, the anionic intermediate **C** is formed, which is a common intermediate for both products **35** and **36**. When **C** is protonated, followed by the elimination of nitrous acid, **35** is obtained (path **a**). The dimeric product **36** is a result of the addition of **C** to another molecule of **16** (path **b**).

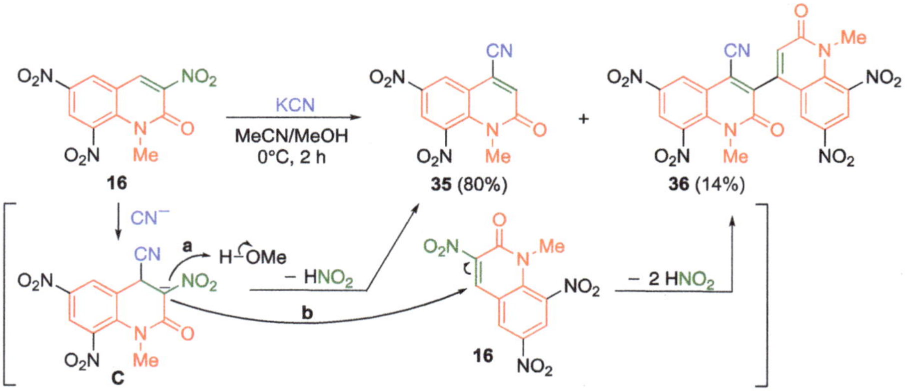

Scheme 11. *cine*-Substitution of **16** with potassium cyanide.

The use of trimethylsilyl cyanide/cesium fluoride instead of potassium cyanide is effective in avoiding undesired dimerization due to the steric hindrance of the *O*-silylated intermediate **D**, affording cyanoquinolone **35** as the sole product without any detectable dimer **36** (Scheme 12). While conventional strategies for cyanation of the **MeQone** framework often involve multistep reactions or harsh conditions, the present method makes the cyanation possible under mild reaction conditions with simple experimental manipulations. Thus, this protocol can be used as a powerful tool for constructing a library of versatile **MeQone** derivatives by further chemical conversion of the cyano and nitro functionalities.

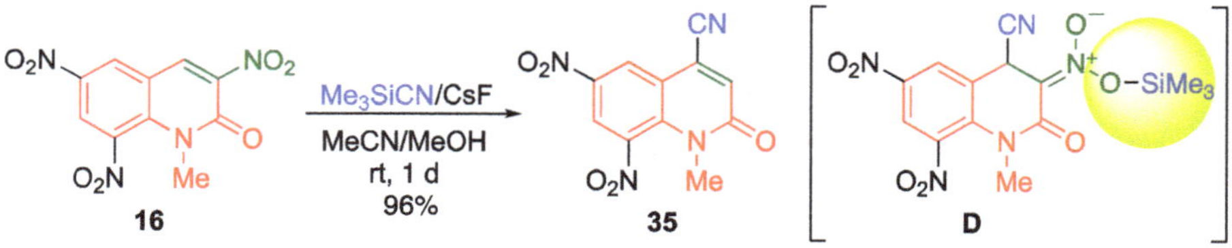

Scheme 12. *cine*-Substitution of **16** with trimethylsilyl cyanide.

The introduction of a methyl group instead of a nitro group at the 8-position also activates the **MeQone** framework. Nitrated 1,8-dimethyl-2-quinolones **37** react with potassium cyanide to afford the corresponding 4-cyano **MeQones** (Table 11).

Table 11. *cine*-Substitution of nitrated 1,8-dimethyl-2-quinolones with trimethylsilyl cyanide.

R⁵	R⁶	R⁷		Yield/%
NO₂	H	NO₂	a	83
NO₂	H	H	b	47
H	NO₂	H	c	quant.

5.1.6. Reaction of Trinitroquinolone with Tertiary Amines

As mentioned in the previous section, the cyanide ion plays two roles: it serves as a nucleophile and it stabilizes anionic intermediate because of its electron-withdrawing nature. Thus, the dimerization of **MeQones** is also observed. Conversely, introduction of a hetero atom at the 4-position generates a stable anionic intermediate, which undergoes efficient dimerization. The treatment of trinitroquinolone **16** with a tertiary amine causes the dimerization [50]. Interestingly, more than two long alkyl chains possessing β-hydrogens are essential for undergoing this reaction (Table 12).

Table 12. Reactions of **16** with tertiary amines.

R¹	R²	R³	Yield/%
Me	Me	Me	0
Et	Et	Et	34
Pr	Pr	Pr	76
Bu	Bu	Bu	93
Bu	Bu	Me	79
Bu	Me	Me	18
PhCH₂	PhCH₂	PhCH₂	0

This reaction is initiated by the nucleophilic addition of tributylamine to trinitroquinolone **16** to produce the zwitterion **E**. The β-elimination of 1-butene is followed by proton transfer of **F** to produce the zwitterion **G**, which reacts with another molecule of **16** to afford dimer **39** (Scheme 13).

Scheme 13. A plausible mechanism for dimerization of **16**.

5.2. Direct C–N Bond Formation at the 4-Position

5.2.1. *cine*-Substitution of Trinitroquinolone with Primary Amines

A different reactivity is observed when primary amines, instead of tertiary amines, are used as the nucleophiles to react with trinitroquinolone **16**. The regioselective C–N bond formation occurs at the 4-position to afford the Meisenheimer complex **40** (Table 13) [51]. When **40** is heated, *cine*-substituted products **41a** and **41b** are obtained; however, no *cine*-substitution is observed for bulky amino substituted derivatives **40c** and **40d**, accompanied by the recovery of considerable amounts of **16**. Upon heating, **40** is converted to dihydroquinolone **H**, from which nitrous acid is eliminated to afford the *cine*-substituted products **41** (Scheme 14, path **a**). However, the elimination of amine proceeds competitively to give the trinitroquinolone **16** (path **b**), which lowers the yield of **41**.

Table 13. *cine*-Substitution of **16** with primary amines.

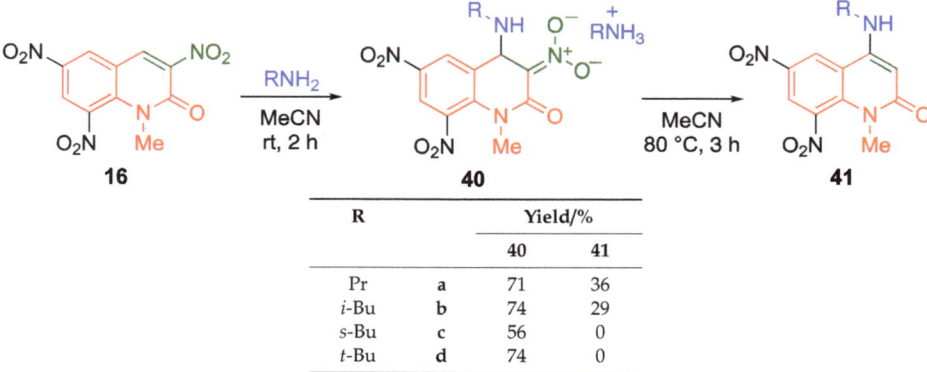

R		Yield/%	
		40	41
Pr	a	71	36
i-Bu	b	74	29
s-Bu	c	56	0
t-Bu	d	74	0

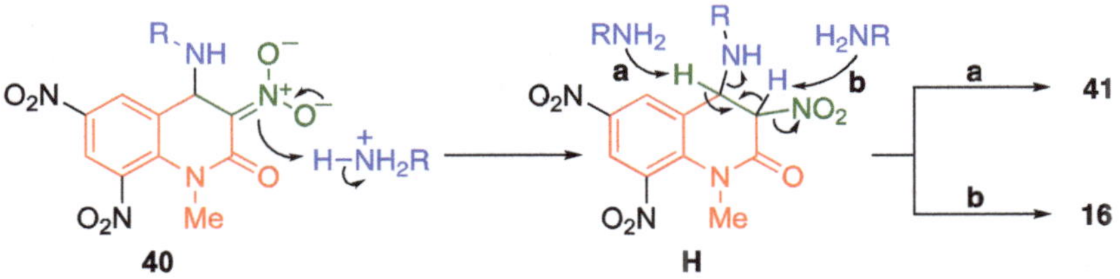

Scheme 14. Two reaction paths leading to 41 and 16.

5.2.2. Amino-Halogenation and Imido-Halogenation of Quinolones

The reaction of trinitroquinolone **16** with excess propylamine in acetonitrile affords the Meisenheimer complex **40a**, which can be used for further functionalization of the **MeQone** framework upon treatment with electrophiles. When the ammonium salt **40a** is treated with N-chlorosuccinimide (NCS), three kinds of functionalized quinolone are obtained; the amino-chlorinated product **42**, the aziridine-fused quinolone **43**, and the imido-chlorinated product **44** (Scheme 15) [52].

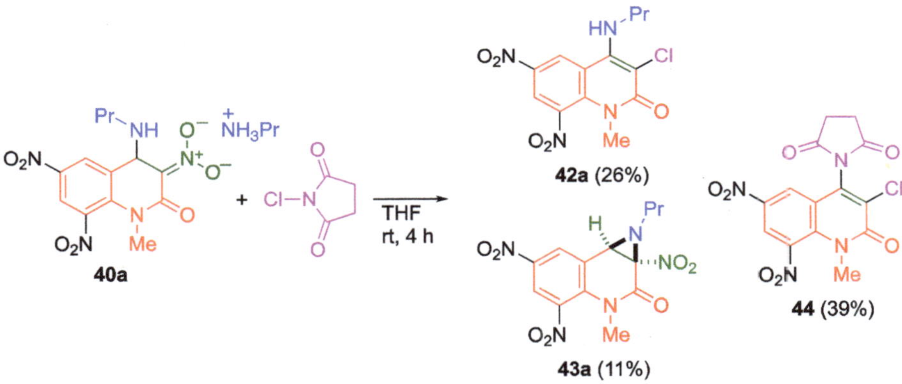

Scheme 15. Reaction of Meisenheimer complex **40a** with NCS.

A plausible mechanism for these reactions is illustrated in Scheme 16. Chlorination of the Meisenheimer complex **40a** affords dihydroquinolone **I**, which is the common intermediate for **42a** and **43a**. The amino-chlorinated product **42a** is formed by elimination of nitrous acid induced by a base, such as imide anion and amine. When the amino group attacks the vicinal position to substitute chloride, an N-propylaziridine ring is formed to give product **43a**. On the other hand, when the eliminated imide anion reacts with trinitroquinolone **16** and NCS, the imido-chlorinated product **44** is formed, which is also formed when **16** is reacted with sodium imide in the presence of NCS (Scheme 17).

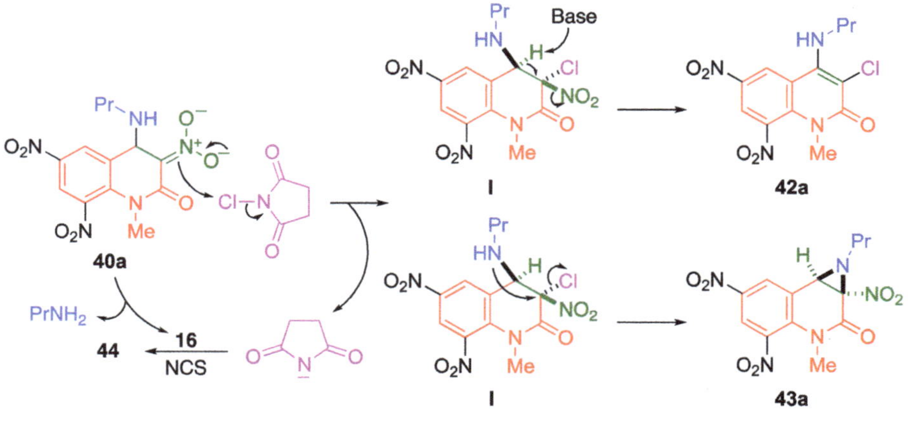

Scheme 16. A plausible mechanism for the formation of **42a** and **43a**.

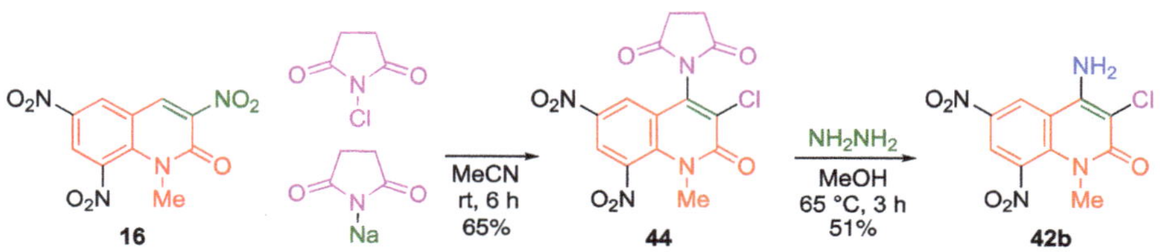

Scheme 17. Synthesis of imido-chlorinated product **44** and the hydrazinolysis.

The amino-halogenation of trinitroquinolone **16** can be conducted in a one-pot two-step manner, in which the selectivity of **42** is increased by using an excess amount of amine (Table 14). The aliphatic and aromatic primary amines undergo the reaction to afford the corresponding amino-chlorinated products **42a–k** in moderate yields. However, less nucleophilic *p*-nitroaniline shows no change. While the acyclic secondary amine, diethylamine, does not furnish **42m**, the cyclic secondary amine, morpholine, yields the corresponding amino-chlorinated product **42n**. Ammonia is difficult to handle in this protocol. Instead, the imido group is considered a masked form of an amino group. Indeed, the imido-chlorinated product **44** can be transformed to the amino-chlorinated quinolone **42b** by hydrazinolysis (Scheme 17).

Table 14. One-pot amino-chlorination of trinitroquinolone **16**.

O_2N-[quinolone **16** with NO_2 at 3-position, O_2N at 6,8-positions, N-Me, C=O] $\xrightarrow[\text{THF, rt, 3 h}]{R^1R^2NH\ (3\ \text{equiv.})}$ $\xrightarrow[\text{THF, rt, 4 h}]{NCS}$ [product **42** with NR^1R^2 at 4-position, Cl at 3-position]

	R^1	R^2		Yield/%
a	Pr	H		62
c	i-Bu	H		70
d	s-Bu	H		49
e	PhCH$_2$	H		54
f	HOCH$_2$CH$_2$	H		56
g	CH$_2$=CHCH$_2$	H		35
h	Ph	H		54
i	4-MeOC$_6$H$_4$	H		37
j	4-BuC$_6$H$_4$	H		41
k	4-IC$_6$H$_4$	H		62
l	4-NO$_2$C$_6$H$_4$	H		trace
m	Et	Et		0
n	-(CH$_2$)$_2$-O-(CH$_2$)$_2$-			62

When NBS is employed as a halogenating reagent, a small amount of the amino-nitrated product **46** is formed in addition to the amino-brominated product **45**, presumably due to the higher leaving ability of bromide than that of chloride (Table 15). Indeed, only amino-nitrated product **46** is obtained without any detectable formation of the iodo-aminated product in the reaction with NIS.

Table 15. Scanning of halogenating agents.

16 $\xrightarrow[\text{THF, rt, 3 h}]{PrNH_2}$ $\xrightarrow[\text{THF, rt, 4 h}]{NXS}$ **45** (NHPr, X) + **46** (NHPr, NO$_2$)

X	Yield/%	
	45	46
Cl	62	0
Br	63	16
I	0	62

5.2.3. Aziridination of Quinolones

The screening of various 3-nitrated **MeQones** reveals the tendency of the selectivity between amino-halogenation and aziridination (Table 16). When the electron density of the benzene ring is low, amino-chlorination occurs predominantly to afford **48a–c**. On the other hand, for increased electron density, intramolecular substitution exclusively occurs to form an aziridine ring, leading to **49d–g**. This tendency is considered to depend on the acidity of the proton at the 4-position in the intermediate **J**. When the acidity of H^4 increases due to the electron-withdrawing group, elimination of a nitrous acid occurs easily via E2 reaction to give the amino-halogenated product **48**. In contrast, when the acidity of H^4 becomes lower, an intramolecular S$_N$2 reaction proceeds to afford the aziridine **49**.

Table 16. Amino-chlorination and aziridination of various 3-nitrated **MeQone**s.

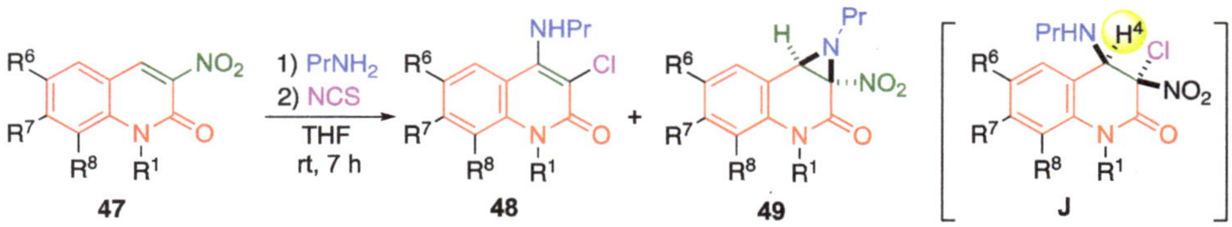

R^1	R^6	R^7	R^8		Yield/%	
					48	49
Me	NO$_2$	H	NO$_2$	a	62	0
Me	NO$_2$	H	Me	b	13	21
Me	NO$_2$	H	H	c	13	49
Me	Br	H	H	d	trace	68
Me	H	H	H	e	0	65
Me	H	H	H	f	0	71
H	H	H	H	g	0	61

1,8-Dimethyl-3,5-dinitro-2-quinolone **50** exhibits a reactivity different from those of the other nitroquinolones **47**. When **50** is subjected to the reaction under the same conditions, *cine*-substitution takes place, rather than amino-chlorination and aziridination, affording compound **51** quantitatively (Scheme 18). In this reaction, the addition of a propylamine affords the Meisenheimer complex **K**. However, the steric repulsion with the *peri*-substituent increases the steric hindrance around the 3-position, thus preventing the attack to NCS. Instead, proton transfer from the 4-position followed by elimination of the nitrite ion affords the *cine*-substituted product **51**.

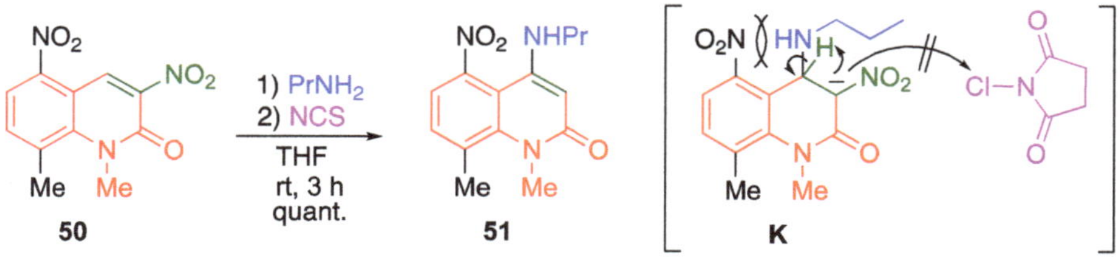

Scheme 18. Different reactivity of 3,5-dinitro-2-quinolone **50** for the amino-chlorination.

Aziridine-fused quinolone **49f** undergoes a ring-opening reaction followed by rearomatization upon treatment with acid, such as toluenesulfonic acid, hydrochloric acid, and trifluoroborane, to furnish the amino-nitrated **MeQone** (Scheme 19).

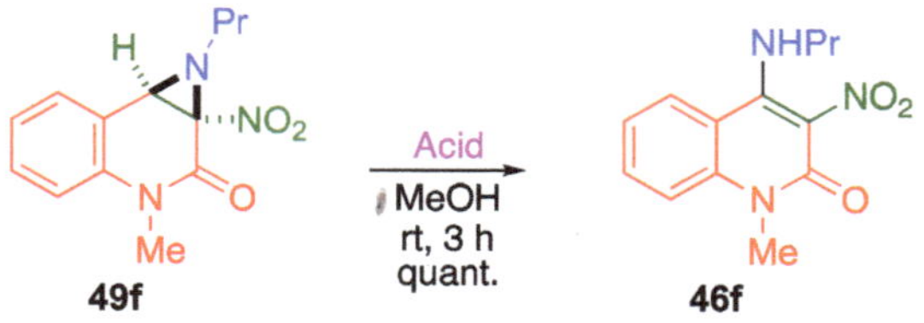

Scheme 19. Aziridine ring opening leading to vicinally functionalized quinolone.

5.3. Direct C–O Bond Formation at the 4-Position

When trinitroquinolone **16** is treated with a sodium alkoxide at room temperature, nucleophilic addition at the 4-position affords an alkoxylated salt **52** [53], which can be isolated because of stabilization by the adjacent nitro and carbonyl groups. After removal of alcohol, treatment of the adduct **52** with NCS in acetonitrile affords the 4-alkoxy-3-chloro-2-quinolone derivatives **53** in moderate-to-high yields (Table 17). This protocol can be performed in a one-pot manner with simple experimental manipulations.

Table 17. Alkoxy-chlorination of **16** leading to **53**.

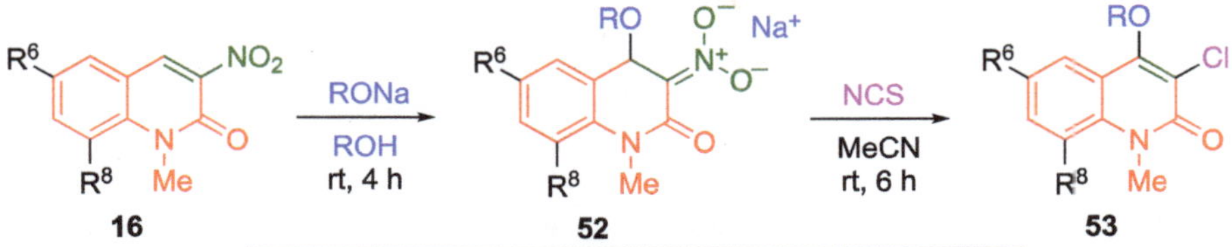

R	R^6	R^8		Yield/%	
				52	53
Me	NO_2	NO_2	a	81	85
Et	NO_2	NO_2	b	quant.	73
i-Bu	NO_2	NO_2	c	-	46
i-Pr	NO_2	NO_2	d	-	45
$PhCH_2CH_2$	NO_2	NO_2	e	-	55
$CH_2=CHCH_2$	NO_2	NO_2	f	-	51
$HC\equiv CCH_2$	NO_2	NO_2	g	-	29
Me	NO_2	Me	h	-	78
Me	NO_2	H	i	-	73
Me	H	H	j	-	65

The reaction proceeds via a similar mechanism, as shown in Scheme 16, for the amino-chlorination (Scheme 20). Chlorination of the alkoxylated salt **52** with NCS affords the dihydroquinolone intermediate **L**, from which nitrous acid is eliminated to form a bis(functionalized) product **53**.

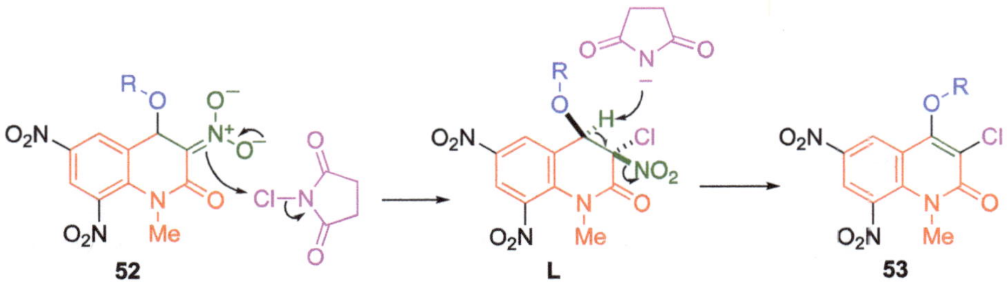

Scheme 20. A plausible mechanism for alkoxy-chlorination of **MeQones**.

When NBS is used as the halogenating reagent, 4-methoxylated trinitroquinolone **54** is obtained in addition to the methoxy-brominated product **53** (Table 18). In the reaction using NIS, product **54** is furnished without any detectable formation of **53**. The different reactivity is due to the higher leaving abilities of bromide and iodide than that of chloride.

Table 18. Scanning of halogenating agents.

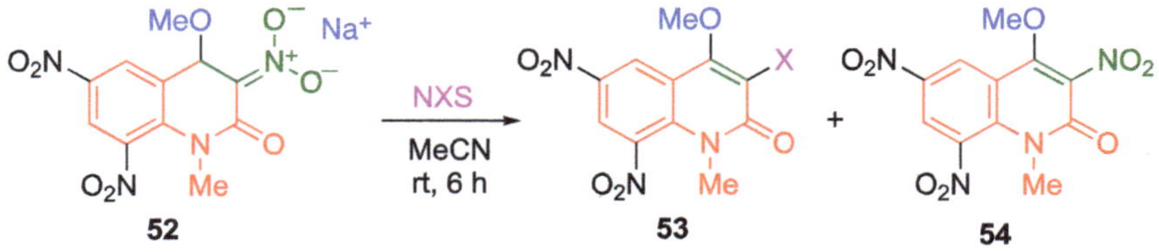

X	Yield/%	
	53	54
Cl	85	0
Br	62	27
I	0	29

3,5-Dinitro-2-quinolone **50** exhibits a reactivity similar to that observed in amino-chlorination to afford the *cine*-substituted product **55** (Scheme 21). Although the addition of methoxide to **50** occurs at the 4-position, it cannot react with NCS at all because of steric repulsion between the 4-methoxy and 5-nitro groups. Instead, proton transfer followed by elimination of nitrite anion affords the *cine*-substituted product **55**.

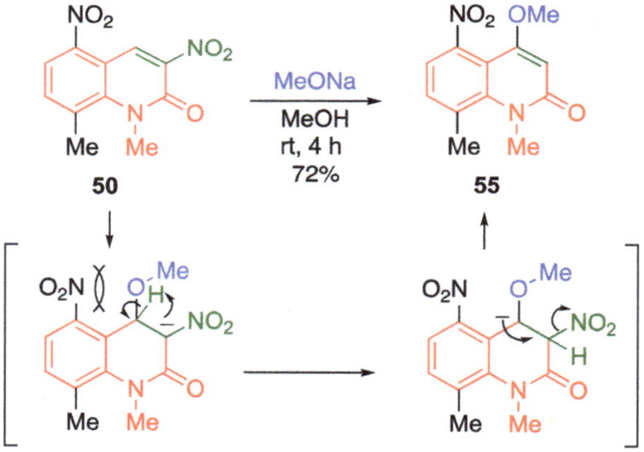

Scheme 21. *cine*-Substitution of 3,5-dinitro-2-quinolone **50** by sodium methoxide.

6. Conclusions

In this review, recent progress in the nitro-promoted direct functionalization of pyridones and quinolones was summarized. A variety of functionalities can be easily introduced into pyridone and quinolone frameworks via activation of the nitro group, facilitating the preparation of newly functionalized derivatives. These methods can promote the construction of a library of pyridones and quinolones with potentially interesting and valuable bioactivities. It is expected that more intensive research in this exciting field will establish the nitro-promoted direct functionalization of heterocycles as a powerful and broadly applicable synthetic strategy in organic synthesis.

Author Contributions: Each author contributed to this article equally. All authors have read and agreed to the published version of the manuscript.

Funding: This research received no external funding.

Conflicts of Interest: The authors declare no conflict of interest.

References

1. Asif, M. A Mini Review: Biological Significances of Nitrogen Hetero Atom Containing Heterocyclic Compounds. *Int. J. Bioorg. Chem.* **2017**, *2*, 146–152.
2. Joule, J.A. Chapter Four-Natural Products Containing Nitrogen Heterocycles-Some Highlights 1990–2015. *Adv. Heterocycl. Chem.* **2016**, *119*, 81–106.
3. Kvasnica, M.; Urban, M.; Dickinson, N.J.; Sarek, J. Pentacyclic Triterpenoids with Nitrogen- and Sulfur-Containing Heterocycles: Synthesis and Medicinal Significance. *Nat. Prod. Rep.* **2015**, *32*, 1303–1330. [CrossRef] [PubMed]
4. Blair, L.M.; Sperry, J. Natural Products Containing a Nitrogen–Nitrogen Bond. *J. Nat. Prod.* **2013**, *76*, 794–812. [CrossRef] [PubMed]
5. Zeidler, J.; Baraniak, D.; Ostrowski, T. Bioactive Nucleoside Analogues Possessing Selected Five-Membered Azaheterocyclic Bases. *Eur. J. Med. Chem.* **2015**, *97*, 409–418. [CrossRef] [PubMed]
6. Hirano, K.; Miura, M. A Lesson for Site-Selective C–H Functionalization on 2-Pyridones: Radical, Organometallic, Directing Group and Steric Controls. *Chem. Sci.* **2018**, *9*, 22–32. [CrossRef]
7. Akihisa, T.; Yokokawa, S.; Ogihara, E.; Matsumoto, M.; Zhang, J.; Kikuchi, T.; Koike, K.; Abe, M. Melanogenesis-Inhibitory and Cytotoxic Activities of Limonoids, Alkaloids, and Phenolic Compounds from Phellodendron amurense Bark. *Chem. Biodiversity* **2017**, *14*, e1700105. [CrossRef]
8. Seya, K.; Miki, I.; Murata, K.; Junke, H.; Motomura, S.; Araki, T.; Itoyama, Y.; Oshima, Y. Pharmacological Properties of Pteleprenine, a Quinoline Alkaloid Extracted from Orixa Japonica, on Guinea-Pig Ileum and Canine Left Atrium. *J. Pharm. Pharmacol.* **1998**, *507*, 803–807. [CrossRef]
9. Kamikawa, T.; Hanaoka, Y.; Fujie, S.; Saito, K.; Yamagiwa, Y.; Fukuhara, K.; Kubo, I. SRS-A Antagonist Pyranoquinolone Alkaloids from East African Fagara Plants and Their Synthesis. *Bioorg. Med. Chem.* **1996**, *4*, 1317–1320. [CrossRef]
10. Joseph, B.; Darro, F.; Béhard, A.; Lesur, B.; Collignon, F.; Decaestecker, C.; Frydman, A.; Guillaumet, G.; Kiss, R. 3-Aryl-2-Quinolone Derivatives: Synthesis and Characterization of In Vitro and In Vivo Antitumor Effects with Emphasis on a New Therapeutical Target Connected with Cell Migration. *J. Med. Chem.* **2002**, *45*, 2543–2555. [CrossRef]
11. Chen, Y.; Wang, F.; Jia, A.; Li, X. Palladium-Catalyzed Selective Oxidative Olefination and Arylation of 2-Pyridones. *Chem. Sci.* **2012**, *3*, 3231–3236. [CrossRef]
12. Nakatani, A.; Hirano, K.; Satoh, T.; Miura, M. Nickel-Catalyzed Direct Alkylation of Heterocycles with α-Bromo Carbonyl Compounds: C3-Selective Functionalization of 2-Pyridones. *Chem. Eur. J.* **2013**, *19*, 7691–7695. [CrossRef] [PubMed]
13. Feng, Z.; Min, Q.; Zhao, H.; Gu, J.; Zhang, X. A General Synthesis of Fluoroalkylated Alkenes by Palladium-Catalyzed Heck-Type Reaction of Fluoroalkyl Bromides. *Angew. Chem. Int. Ed.* **2014**, *53*, 1–6.
14. Nakatani, A.; Hirano, K.; Satoh, T.; Miura, M. Manganese-Mediated C3-Selective Direct Alkylation and Arylation of 2-Pyridones with Diethyl Malonates and Arylboronic Acids. *J. Org. Chem.* **2014**, *79*, 1377–1385. [CrossRef] [PubMed]
15. Elenicha, O.V.; Lytvyn, R.Z.; Skripskaya, O.V.; Lyavinets, O.S.; Pitkovych, K.E.; Yagodinets, P.I.; Obushak, M.D. Synthesis of Nitrogen-Containing Heterocycles on the Basis of 3-(4-Acetylphenyl)-1-methylquinolin-2(1H)-one. *Russ. J. Org. Chem.* **2016**, *52*, 373–378. [CrossRef]
16. Li, J.; Hu, D.; Liang, X.; Wang, Y.; Wang, H.; Pan, Y. Praseodymium(III)-Catalyzed Regioselective Synthesis of C_3-N-Substituted Coumarins with Coumarins and Azides. *J. Org. Chem.* **2017**, *82*, 9006–9011. [CrossRef]
17. Prendergast, A.M.; McGlacken, G.P. Transition Metal Mediated C–H Activation of 2-Pyrones, 2-Pyridones, 2-Coumarins and 2-Quinolones. *Eur. J. Org. Chem.* **2018**, 6068–6082. [CrossRef]
18. Maity, S.; Das, D.; Sarkar, S.; Samanta, R. Direct Pd(II)-Catalyzed Site-Selective C5-Arylation of 2-Pyridone Using Aryl Iodides. *Org. Lett.* **2018**, *20*, 5167–5171. [CrossRef]
19. Diesel, J.; Finogenova, A.M.; Cramer, N. Nickel-Catalyzed Enantioselective Pyridone C–H Functionalizations Enabled by a Bulky N-Heterocyclic Carbene Ligand. *J. Am. Chem. Soc.* **2018**, *140*, 4489–4495. [CrossRef]
20. Hazra, S.; Hirano, K.; Miura, M. Solvent-Controlled Rhodium-Catalyzed C6-Selective C–H Alkenylation and Alkynylation of 2-Pyridones with Acrylates. *Asian J. Org. Chem.* **2019**, *8*, 1097–1101. [CrossRef]

21. Zhao, H.; Xu, X.; Luo, Z.; Cao, L.; Li, B.; Li, H.; Xu, L.; Fan, Q.; Walsh, P.J. Rhodium(I)-Catalyzed C6-Selective C–H Alkenylation and Polyenylation of 2-Pyridones with Alkenyl and Conjugated Polyenyl Carboxylic Acids. *Chem. Sci.* **2019**, *10*, 10089–10096. [CrossRef]
22. Nishiwaki, N. Chemistry of Nitroquinolones and Synthetic Application to Unnatural 1-Methyl-2-quinolone Derivatives. *Molecules* **2010**, *15*, 5174–5195. [CrossRef] [PubMed]
23. Calderari, G.; Seebach, D. Asymmetric *Michael*-Additions. Stereoselective Alkylation of Chiral, Non-racemic Enolates by Nitroolefins. Preparation of Enantiomerically Pure γ-Aminobutyric and Succinic Acid Derivatives. *Helv. Chim. Acta* **1985**, *68*, 1592–1604. [CrossRef]
24. Ballini, R.; Bosica, G.; Fiorini, D.; Petrini, M. Unprecedented, Selective Nef Reaction of Secondary Nitroalkanes Promoted by DBU under Basic Homogeneous Conditions. *Tetrahedron Lett.* **2002**, *43*, 5233–5235. [CrossRef]
25. Wehrli, P.A.; Schaer, B. Direct Transformation of Primary Nitro Compounds into Nitriles. New Syntheses of α,β-Unsaturated Nitriles and Cyanohydrin Acetates. *J. Org. Chem.* **1977**, *42*, 3956–3958. [CrossRef]
26. Orlandi, M.; Brenna, D.; Harms, R.; Jost, S.; Benaglia, M. Recent Developments in the Reduction of Aromatic and Aliphatic Nitro Compounds to Amines. *Org. Process Res. Dev.* **2018**, *22*, 430–445. [CrossRef]
27. Yadav, M.R.; Nagaoka, M.; Kashihara, M.; Zhong, R.-L.; Miyazaki, T.; Sakai, S.; Nakao, Y. The Suzuki–Miyaura Coupling of Nitroarenes. *J. Am. Chem. Soc.* **2017**, *139*, 9423–9426. [CrossRef]
28. Inoue, F.; Kashihara, M.; Yadav, M.R.; Nakao, Y. Buchwald–Hartwig Amination of Nitroarenes. *Angew. Chem. Int. Ed.* **2017**, *56*, 13307–13309. [CrossRef]
29. Halimehjani, A.Z.; Namboothiri, I.N.N.; Hooshmand, S.E. Nitroalkenes in the Synthesis of Carbocyclic Compounds. *RSC Adv.* **2014**, *4*, 31261–31299. [CrossRef]
30. Hao, F.; Nishiwaki, N. Chemistry of Nitroaziridines. *Heterocycles* **2019**, *99*, 54–72.
31. Hao, F.; Yokoyama, S.; Nishiwaki, N. Direct Dihalo-Alkoxylation of Nitroalkenes Leading to β,β-Dihalo-β-nitroethyl Alkyl Ethers. *Org. Biomol. Chem.* **2018**, *16*, 2768–2775. [CrossRef] [PubMed]
32. Asahara, H.; Sofue, A.; Kuroda, Y.; Nishiwaki, N. Alkynylation and Cyanation of Alkenes Using Diverse Properties of a Nitro Group. *J. Org. Chem.* **2018**, *83*, 13691–13699. [CrossRef] [PubMed]
33. Andreassen, E.J.; Bakke, J.M.; Sletvold, I.; Svensen, H. Nucleophilic Alkylations of 3-Nitropyridines. *Org. Biomol. Chem.* **2004**, *2*, 2671–2676. [CrossRef] [PubMed]
34. Fujita, R.; Watanabe, K.; Nishiuchi, Y.; Honda, R.; Matsuzaki, H.; Hongo, H. Diels-Alder Reactions of Nitro-2(1*H*)-pyridones with 2,3-Dimethyl-1,3-butadiene. *Chem. Pharm. Bull.* **2001**, *49*, 601–605. [CrossRef] [PubMed]
35. Fujita, R.; Yoshisuji, T.; Wakayanagi, S.; Wakamatsu, H.; Matsuzaki, H. Synthesis of 5(6*H*)-Phenanthridones Using Diels-Alder Reaction of 3-Nitro-2(1*H*)-quinolones Acting Dienophiles. *Chem. Pharm. Bull.* **2006**, *54*, 204–208. [CrossRef] [PubMed]
36. Fujita, R.; Watanabe, K.; Yoshisuji, T.; Kabuto, C.; Matsuzaki, H.; Hongo, H. Diels-Alder Reaction of 2(1*H*)-quinolones Having an Electron-Withdrawing Group at the 3-Position Acting as Dienophiles with Dienes. *Chem. Pharm. Bull.* **2001**, *49*, 893–899. [CrossRef]
37. Asahara, M.; Nagamatsu, M.; Tohda, Y.; Nishiwaki, N.; Ariga, M. Diels-Alder Reaction of 1-Methyl-3,6,8-trinitro-2-quinolone. *J. Heterocycl. Chem.* **2004**, *41*, 803–805. [CrossRef]
38. Asahara, M.; Shibano, C.; Koyama, K.; Tamura, M.; Tohda, Y.; Ariga, M. The Nitroalkene Showing Dual Behaviors in the Same Reaction System. *Tetrahedron Lett.* **2005**, *46*, 7519–7521. [CrossRef]
39. Nishiwaki, N.; Tanaka, C.; Asahara, M.; Asaka, N.; Tohda, Y.; Ariga, M. A Nitro Group Distorting 2-Quinolone Skeleton. *Heterocycles* **1999**, *51*, 567–574.
40. Murashima, T.; Nishi, K.; Nakamoto, K.; Kato, A.; Tamai, R.; Uno, H.; Ono, N. Preparation of Novel Heteroisoindoles from Nitropyridines and Nitropyridones. *Heterocycles* **2002**, *58*, 301–310. [CrossRef]
41. Nishiwaki, N.; Tanaka, A.; Uchida, M.; Tohda, Y.; Ariga, M. *cine*-Substitution of 1-Methyl-3,6,8-trinitro-2-quinolone. *Bull. Chem. Soc. Jpn.* **1996**, *69*, 1337–1381. [CrossRef]
42. Chen, X.; Kobiro, K.; Asahara, H.; Kakiuchi, K.; Sugimoto, R.; Saigo, K.; Nishiwaki, N. Reactive 2-Quinolones Dearomatized by Steric Repulsion between 1-Methyl and 8-Substituted Groups. *Tetrahedron* **2013**, *69*, 4624–4630. [CrossRef]
43. Asahara, M.; Ohtsutsumi, M.; Ariga, M.; Nishiwaki, N. Regioselective Nitroalkylation of the 1-Methyl-2-quinolone Framework. *Heterocycles* **2009**, *78*, 2851–2854.
44. Asahara, M.; Katayama, T.; Tohda, Y.; Nishiwaki, N.; Ariga, M. Synthesis of Unnatural 1-Methyl-2-quinolone Derivatives. *Chem. Pharm. Bull.* **2004**, *52*, 1134–1138. [CrossRef] [PubMed]

45. Aasahara, M.; Ohtsutsumi, M.; Tamura, M.; Nishiwaki, N.; Ariga, M. Electrophilic Arylation of Phenols: Construction of a New Family of 1-Methyl-2-quinolones. *Bull. Chem. Soc. Jpn.* **2005**, *78*, 2235–2237. [CrossRef]
46. Fleming, F.F.; Yao, L.; Ravikumar, P.C.; Funk, L.; Shook, B.C. Nitrile-Containing Pharmaceuticals: Efficacious Roles of the Nitrile Pharmacophore. *J. Med. Chem.* **2010**, *53*, 7902–7917. [CrossRef] [PubMed]
47. Zhu, Y.; Li, Y.; Xiang, S.; Fan, W.; Jin, J.; Huang, D. Utilization of Nitriles as the Nitrogen Source: Practical and Economical Construction of 4-Aminopyrimidine and β-Enaminonitrile Skeletons. *Org. Chem. Front.* **2019**, *6*, 3071–3077. [CrossRef]
48. Ghosh, T.; Si, A.; Misra, A.K. Facile Transformation of Nitriles into Thioamides: Application to C-Glycosyl Nitrile Derivatives. *ChemistrySelect* **2017**, *2*, 1366–1369. [CrossRef]
49. Xi, F.; Kamal, F.; Schenerman, M.A. A Novel and Convenient Transformation of Nitriles to Aldehydes. *Tetrahedron Lett.* **2002**, *43*, 1395–1396. [CrossRef]
50. Nishiwaki, N.; Sakashita, M.; Azuma, M.; Tanaka, C.; Tamura, M.; Asaka, N.; Hori, K.; Tohda, Y.; Ariga, M. Novel Functionalization of 1-Methyl-2-quinolone; Dimerization and Denitration of Trinitroquinolone. *Tetrahedron* **2002**, *58*, 473–478. [CrossRef]
51. Asahara, M.; Nagamatsu, M.; Tohda, Y.; Nishiwaki, N.; Ariga, M. Effective C-N Bond Formation on the 1-Methyl-2-quinolone Skeleton. *ARKIVOC* **2005**, *1*, 1–6.
52. Hao, F.; Asahara, H.; Nishiwaki, N. Direct Amino-halogenation and Aziridination of the 2-Quinolone Framework by Sequential Treatment of 3-Nitro-2-quinolone with Amine and *N*-Halosuccinimide. *Tetrahedron* **2017**, *73*, 1255–1264. [CrossRef]
53. Hao, F.; Asahara, H.; Nishiwaki, N. A Direct and Vicinal Functionalization of the 1-Methyl-2-quinolone Framework: 4-Alkoxylation and 3-Chlorination. *Org. Biomol. Chem.* **2016**, *14*, 5128–5135. [CrossRef] [PubMed]

© 2020 by the authors. Licensee MDPI, Basel, Switzerland. This article is an open access article distributed under the terms and conditions of the Creative Commons Attribution (CC BY) license (http://creativecommons.org/licenses/by/4.0/).

MDPI
St. Alban-Anlage 66
4052 Basel
Switzerland
Tel. +41 61 683 77 34
Fax +41 61 302 89 18
www.mdpi.com

Molecules Editorial Office
E-mail: molecules@mdpi.com
www.mdpi.com/journal/molecules

www.ingramcontent.com/pod-product-compliance
Lightning Source LLC
LaVergne TN
LVHW070044120526
838202LV00101B/429